Technology in the Industrial Revolution

Technological change is about more than inventions. This concise history of the Industrial Revolution places the eighteenth-century British Industrial Revolution in global context, locating its causes in government protection, global competition, and colonialism. Inventions from spinning jennies to steam engines came to define an age that culminated in the acceleration of the fashion cycle, the intensification in demand and supply of raw materials and the rise of a plantation system that would reconfigure world history in favor of British (and European) global domination. In this accessible analysis of the classic case of rapid and revolutionary technological change, Barbara Hahn takes readers from the north of England to slavery, cotton plantations, the Anglo-Indian trade and beyond – placing technological change at the center of world history.

Barbara Hahn is a prize-winning author in business history and the history of technology. Her publications include *Plantation Kingdom: The South and Its Global Commodities* (2016), which she co-authored. She is Associate Professor of History at Texas Tech University and was the associate editor of the journal *Technology and Culture*.

New Approaches to the History of Science and Medicine

This dynamic new series publishes concise but authoritative surveys on the key themes and problems in the history of science and medicine. Books in the series are written by established scholars at a level and length accessible to students and general readers, introducing and engaging major questions of historical analysis and debate.

Other Books in the Series

Technology in the Industrial Revolution

Barbara Hahn
Texas Tech University

CAMBRIDGE
UNIVERSITY PRESS

CAMBRIDGE
UNIVERSITY PRESS

University Printing House, Cambridge CB2 8BS, United Kingdom

One Liberty Plaza, 20th Floor, New York, NY 10006, USA

477 Williamstown Road, Port Melbourne, VIC 3207, Australia

314–321, 3rd Floor, Plot 3, Splendor Forum, Jasola District Centre,
New Delhi – 110025, India

79 Anson Road, #06–04/06, Singapore 079906

Cambridge University Press is part of the University of Cambridge.

It furthers the University's mission by disseminating knowledge in the pursuit of
education, learning, and research at the highest international levels of excellence.

www.cambridge.org
Information on this title: www.cambridge.org/9781107186804
DOI: 10.1017/9781316900864

First published 2020

A catalogue record for this publication is available from the British Library.

ISBN 978-1-107-18680-4 Hardback
ISBN 978-1-316-63746-3 Paperback

For the British working people

Contents

Figures

Acknowledgments

This book is a product of a long conversation with people, places, and primary sources, both documents and artifacts. I first went to England in 1989, and worked as a cleaner in a Pontin's on Camber Sands. I returned to investigate this topic in 2012, as a guest of the Marcus Cunliffe Centre for the Study of the American South, at the University of Sussex, where I spoke about tobacco. There was then a mad dash through the industrial museums of the North based around a very comfortable ten-day stay in Burnley. I thank my guides over the years and their institutions: the downloadable walking tour of industrial Manchester offered by the Ancoats History Project, Bolton Library and Museum Services, the Bradford Industrial Museum, Bridport Museum, Crewkerne and District Museum, The Dean Clough Mills, Helmshore Mills Textile Museum, Kelvingrove Art Gallery and Museum, Leeds City Museum, Leeds Discovery Centre, Leeds Industrial Museum at Armley Mills, Museo della Setta in Como, Manchester Museum of Science and Industry, New Lanark Village, Quarry Bank Mill in Styal, Queen Street Mill in Harle Syke, Paisley Museum and Art Galleries, the People's History Museum, Saltaire, Stockport Hat Museum, Temple Newsam, Thwaite Mills Watermill, a twine works ropewalk, Verdant Works jute mill in Dundee, the Westbury Manor Museum in Fareham, the Whitchurch Silk Mill, Winchester City Flour Mill, and the Working Class Movement Library in Salford. I learned so much there, and I am so thankful for the opportunity. Some of these have since closed.

In summer 2013, I was back again, a month in Liverpool and two in Manchester, supported by Texas Tech University (thank you!), and then was lucky to spend two years in Leeds as a Marie Curie International Incoming Fellow (MCIIF #628722) of the European Commission, as part of "Rethinking Textiles," (itself part of the "Enterprise of Culture" project), whose principal investigator was Regina Lee Blaszczyk. Special thanks to Reggie Blaszczyk, the European Union, the University of Leeds, and swathes of its staff, especially Mike Bellhouse, Fiona Blair, and Paul

McShane – and, quite separately, my Leeds landlord, Mr. John S. House, of St. John's Terrace, at the top of the Bellevue Road.

Immersion was invaluable. It helped me understand Britain better than I had done before: the rock-bottom importance of class structure to the society made there and contested there every day, in every way – in the fourteenth and the eighteenth centuries as well as today. I began to realize that class is as important to English history as race is to American history. Being there over time also introduced me to a large and diverse community of scholars and other people talking and thinking about the Industrial Revolution. My American Midwestern tendency to talk to strangers could open up conversations, especially in the North, and many people I encountered offered their own interpretations, shared local and family history, or suggested places to visit or things to read. I followed my nose. Halifax and Hebden Bridge, Huddersfield and even Honley, Preston and Shipley and Sheffield, Hawick and Jedburgh, Glasgow and Edinburgh, and, of course, Manchester and Liverpool and Leeds, the industrial North and the Scottish borders. At one point, outdoors on Deansgate, in Manchester, over a bowl of soup, John Pickstone asked, "Who have you spoken to?" Invitations flowed from there. Some of these people I already knew and some came from Pickstone and others from others. I wish I could explain to you how each one helped, or tell a story about each one so you'd know how amazing they are:

Larra Anderson, Robert J. Aram, William Ashworth, Kellen Backer, Sharon Bainbridge, Bruce E. Baker, Sara Barker, Sarah Barton, Maxine Berg, Leonie Betts, Linda Betts, Fiona Blair, Regina Lee Blaszczyk, Tilly Blyth, Kathryn Boodry, Anna Bowman, Emily Buchnea, Robert Bud, Sarah Butler, Malcolm Chase, Eric Chiappinelli, David Churchill, Peter Coclanis, Gill Cookson, Hansa and Kish Dabhi, Keith Dando, Sarah Dietz, Alice Dolan, Peter Doré, Kate Dossett, Robert Du Plessis, Gökhan Ersan, Elaine Evans, Helen Farrar, Mike Finn, Richard Follett, Ester Galeci, Graeme Gooday, William Gould, Shane Hamilton, Sasha Handley, Michael Hann, James Harris, Abigail Harrison-Moore, Tiana Harper, Jan Hersheimer, Jan Hicks, Richard High, the Reverend Richard L Hills, Steve Hindle, Philip T. Hoffman, Pat Hudson, Jane Humphries, Karolina Hutkova, Kenneth Jackson, Amy Jenkinson, Finn Arne and Dolly and Marion and Lina Jørgensen, Jen Kaines, Nina Kane, Hannah Kemp, Jack Kirby, Ursula Klein, Kazuo Kobayashi, Renée Lane, David Larmour, Mitch Larson, Pamela Long, Graham Loud, Antonia Lovelace, Andreas Malm, Annapurna Mamidipudi, Daniel Martin, Judith Mary Martin and Nigel Martin, Peter Maw, Meg McHugh, Liz McIvor, Vincent McKernan, Philippe Minard, Lesley Miller, Laura Millward, Luca Mola, Craig Muldrew, Anne Murphy, Camilla Nichol, Eugene Nicholson, Tom Nies, Scot Ninnemann, Lisa O'Brien, Nicholas Oddy, Alan Olmstead, Jenifer Parks, Prasannan Parthasarathi, Bruce Peter, John Pickstone, Andrew Popp, Paul and Mindy and Alice Quigley, Anita Quye, Greg Radick, Natalie Raw, Phil Reekers, Paul Rhode, Giorgio Riello, Lissa

Roberts, Frances Robertson, Alex Roland, Stephanie Roper, Mary B. Rose, Jean-Laurent Rosenthal, Sally Rush, Ann Schofield, Yda Schreuder, Jan Shearsmith, Pamela Smith, Jennifer Snead, Danielle Sprecher, John Styles, Keith Sugden, Ann Sumner, James Sumner, Abigail Swingen, Philip Sykas, Fiona Tait, Steve Toms, E. B. Toon, Sally Tuckett, Leucha Veneer, Jeff Waddington, Andrew Walden, Claire Watson, Dan Weldon, Corbin Williamson, Susan Williamson, Deirdre and Nigel Wood, Chris Wrigley, Joseph Wright-Pangolin, and last but not least – Natalie Zacek.

There are so many more who helped, whose names I don't know: archivists, tour guides, the barmaid who worked the locks on the Manchester canal tour (twice). Likewise thanks to the kind woman who delayed her lunch to give me an unscheduled tour of Johnston's cashmere mill in Hawick. She picked up a jumper from her chair and carried it through the factory and showed me what each machine did in putting it together, and for the first time I understood machine knitting a little bit. Sasha Handley, Elizabeth Toon, and Natalie Zacek opened their homes to me for extended stays. James Sumner and Leucha Veneer, and Nigel and Judith Mary Martin, provided tours and discussions of specific regions, and ropewalks, and the sailcloth industry. I'm sad that I could not include every detail of what each one helped me learn. Between book contract and due date I also benefitted tremendously from a month at the Max-Planck-Institut für Wissenschaftsgeschichte in Berlin, where chapters four and five first took shape and where the Moving Crops collaborators kept me focused on assemblages in the form of crops-capes. Thanks to Dagmar Schäffer and her office angels, and Francesca Bray, Alina-Sandra Cucu, John Bosco Lourdusamy, and Tiago Saraiva. Similar quiet hallways, library service, and lunch, permeated a summer residency at the National Humanities Center, where I am grateful to have had the opportunity to begin revising my first draft.

Gill Cookson, Eugene Nicholson, and Alex Roland deserve thanks for reading a first version of the entire manuscript and saving me from significant errors of fact, tone, and interpretation. Thanks too to Lucy Rhymer, and the anonymous readers whose comments changed my course midway through. Audiences and their questions also helped me think through this project. Over the years, they have included: the Arkwright Society, Association of Business Historians, the Business History Conference, Caltech and the Huntington Library, the Centre for the History and Philosophy of Science at Leeds, the Glasgow School of Art (Focus on Critical Inquiry), Heritage Show+Tell, ICHOTEC and ICHSTM, the Institute for the Study of Western Civilization at Texas Tech, the Society for the History of Technology (SHOT), Umeå University, University of Liverpool (Port City Lives), and the University of York Management School.

Flaws that remain are, of course, my own.

Texas Tech has been my intellectual home for more than a decade, and I am grateful for the support and friendships nourished here. The indefatigable Document Delivery office and Jack Becker in the library deserve special thanks. The TTU Humanities Center subvented the publication of illustrations. Sean Cunningham and Randy McBee helped me meet ambitious research goals. I thank them and the whole institution.

Lancashire and Yorkshire and the people of the British Isles have themselves taught me so much. It has been a pleasure to be your guest.

Introduction

The Industrial Revolution was about more than inventions. Instead of individual machines conceived by heroic inventors in a flash of discovery, this is a book about systems and networks, and the worlds that got the machines running, and the way the world changed to make the devices work. Making machinery operational required resources only available outside the machines themselves, even as getting the machines running meant rearrangements of labor and power and raw material supplies, of markets and distribution and finance, and of consumer tastes and global geopolitics. The famous machines used inputs to operate. They also needed markets for the things they made, and marketing too. Without demand, without buyers for their products, mechanization would have failed because investing capital in machines would not have made profits. Without supplies and without workers, likewise, the storied inventions of the Industrial Revolution would have stopped. The chimneys and mills, the wheels and water and feudal arrangements for their use, slave plantations and guild prerogatives, transportation and communication networks, government protection and imperial competition, all played a role in making specific devices work. The relationship between changing machines and their changing contexts is the subject this book investigates.

The classic case of Industrial Revolution refers to transformations in the textile business of northern Britain in the late eighteenth century. These changes included using machines, powered by inanimate energy sources, to make cloth in factories, resulting in mass production of fabric that sold around the globe. These shifts were a response: the reaction of regional cloth merchants to their fears about cotton fabrics imported from India. These clothiers operated a dynamic, powerful, intricately articulated textile industry that had taken shape in medieval England before expanding in the 1500s and 1600s with the new commercial links between Europe and the trade of the Indian Ocean. In the eighteenth century, merchants who responded to these changes by investing capital in factories and machinery became industrial capitalists – a new kind of businessman. They did not act alone, though. Britain's tradesmen

1

became industrial manufacturers with the help of the state. It was not laissez-faire capitalism that made industrialization possible: the mercantilist ideology of the age ensured government support for both international adventures and home industry. Instead, a protected domestic market and competition for consumers elsewhere – in Africa and America, especially – provided the economic context for the piecemeal adoption and combination of practices and devices that worked, over time, as the new emerging technologies of the day. Social structures and existing institutions, from market rights to the established Church of England, influenced when and how machinery worked in everyday but changing systems of production.

Justification

The Industrial Revolution as a coherent episode of technological change is a bit of a conceit, constructed by later minds out of a wide array of incidents that seemed fairly random at the time. The usual events evoked by the phrase "Industrial Revolution" took place in a few counties in the British North, between the 1760s and the 1840s, where the local cloth merchants of a national textile industry were changing how they did business. Within shifting frameworks of government protection and global competition, they invested in machinery to ensure regular supplies of goods to sell. They contributed to reorganizing global trade patterns, individual household structures, and consumer fashion as part of that process.

The revolutionary nature of their activities can be established by certain data, which indicate what this book attempts to explain: how did cotton manufacturing increase so dramatically in the north of England in the 1760s to 1840s? What role did machinery play? Where did the devices come from, and how were they made to work? What changed, and what stayed the same, in British cloth production in those eight decades? Answering these questions means seeing machinery as the heart of the story and attempting to understand its origins as well as its impact. Placing machinery in a middle position, as both cause and effect, provides frameworks to explain the rise of mass production, one result of industrialization. Mass consumption and waged labor were additional results of industrialization, as was a specific imperialist organization of the globe. The data on cotton consumption, on the value of output of the cotton industry, and on the importance of cotton to the larger British economy, therefore voice the questions that this book attempts to answer.

Manufacturing means turning unrefined commodities into consumer goods, so the consumption of commodified raw material is one thing this book wishes to explain. Raw cotton imports to Britain increased a hundredfold during the Industrial Revolution, from 4.2 million pounds in 1772 to 41.8 million in 1800 and 452 million in 1841. In terms of the output: in 1766, Britain exported cotton goods worth a total of £221,000; in 1800 the figure was £5.4 million (two-fifths of the nation's total export of manufactured goods). By 1840, the figure was £22 million. (This would be another hundredfold increase if currency measures were not often distorted by inflation). Cotton was 1 percent of British industry in 1770, rising to 10 percent of the whole manufacturing sector of the British Isles in 1841.[1] In comparison, the tech sector was 0.8 percent of US Gross Domestic Product in 1980 and climbed to 5.2 percent by 2015.[2] The cotton industry of Britain rose higher, though not as fast. By the end of our story, cotton manufacturing was a more important part of the British economy than the tech industry is for the United States today.

These data indicate a shift in human experience. The rapid increase in the inputs and outputs of cotton, in a few counties in the north of Britain, about 150 years ago, accompanied a fundamental change in how people live. It was in this period that most people changed from making what they used to buying those things that other people made. Very few of us today spin our own yarn, weave our own cloth, or sew our own clothing. Most of what we wear was made by other people, far away, on machines owned by yet others; in other words, was made industrially. The end of this story, the mass production of textiles for distant consumers, had its beginnings in a very different set of goals. Men who bought machinery to multiply and improve on traditional work sought to imitate luxurious imported fabrics. They also pursued and achieved government protection for their industry, and they sold their goods into colonies and around the world. During the Industrial Revolution, technological systems around the adoption of new machinery shifted from product innovation (making new goods) to cost competition (making goods more cheaply), from luxury and niche production to mass market and mass production. The individual person's experience of making and using goods is different, as a result, than it was before industrialization.

---

[1] Joseph Inikori, *Africans and the Industrial Revolution in England: A Study in International Trade and Economic Development* (Cambridge and New York: Cambridge University Press, 2002), 78–79; David S. Landes, *Unbound Prometheus: Technological Change and Industrial Development in Western Europe from 1750 to the Present* (Cambridge: Cambridge University Press, 1969), 41–42; Edward Baines, Jr., *History of the Cotton Manufacture in Great Britain* (London: H. Fisher, R. Fisher, and P. Jackson, 1835), 215; Christine MacLeod, *Heroes of Invention: Technology, Liberalism, and British Identity, 1750–1914* (Cambridge and New York: Cambridge University Press, 2007), 64; "Commercial Statistics: Annual Export of British Manufactures," *Hunt's Merchant's Magazine* 5 (July 1841), 385.

[2] Prasannan Parthasarathi, *Why Europe Grew Rich and Asia Did Not: Global Economic Divergence, 1600–1850* (Cambridge and New York: Cambridge University Press, 2011), 12; Ian Hathaway, "How Big is the Tech Sector?," blogpost, May 31, 2017.

Definitions

A three-part definition of industrialization guides the analysis in this book:

1. The mechanization of some task or series of tasks previously done by hand. New machinery is often taken as the cause of technological change. Here it is examined as a physical expression of the changing world around it. Both the development and adoption of machinery are part of a bigger story about how methods changed, and who did the work, and where the raw materials came from and where the finished goods went.

2. The separation of production from consumption – the people who use goods are not the ones who made them. This often includes the removal of work from home, and of labor from leisure. It also intimates a division of labor, in which the tasks of making something are performed by different people, instead of by one person who makes a piece from beginning to end.

3. The development of regular flow and standard characteristics of physical objects, from raw materials to finished products, from supply to demand. This incorporates the predictable procurement of expected materials, and established links to distribute the finished goods. One example is the cotton trade between North America and Britain that fed raw materials into machines in factories as these developed. The Industrial Revolution both relied upon and stimulated this regular flow of materials. Another example is consistency of goods mass-produced, rather than made individually.

This three-part definition draws on a similar description, given in the first pages of David Landes' 1969 book *Unbound Prometheus*, which defined industrialization primarily in terms of the replacement of muscle with inanimate power. This emphasis on generalized inanimate power neglects important differences in the technological systems around each type – wind, water, steam – as well as the role of horses (muscle power) in early industries. Treating inanimate power as one single thing therefore omits the contingencies that made one system more attractive than another in specific times and places. So instead "mechanization" here replaces that part of the definition, and the energy source that powered a machine will be treated as part of its production system. The second part of the definition draws on recent scholarly recognition that changes in consumption accompany or even lead to changes in production – that demand helps explain changes in supply. Finally, the third part of this definition expands on Landes's "marked improvement in the getting and working of raw materials, especially in what are now known as the metal-lurgical and chemical industries." By keeping the focus of this book on one single sector, here the transformations in raw material supplies include the use and technological expansion of the plantation system of production to the cultivation of cotton in the American slave south.[3]

Existential Questions

Did the Industrial Revolution really happen? Scholars ask: Was it a revolution – a dramatic change in a short period of time? Was it industrial, or did another sector (agriculture, for example) experience greater efficiency gains? Those who dispute the idea of an Industrial Revolution have good evidence on their side. They point out the component mechanization, stretching back into medieval wool finishing and silk throwing, that this book sketches in Chapter 1. Much of the machinery so often used to explain technological change was already available, and some was even adopted, in the long centuries before the few decades we call the Industrial Revolution. Quantitatively, too, the late eighteenth-century "spurt" of manufacturing output "was confined largely to cotton goods" and "it was not until the 1820s that the quantitative weight of new industries imposed itself on the economy as a whole." Change was gradual, investment small relative to the entire economy, and only a small percentage of even the English population experienced the dislocations and opportunities of the age.[4]

Contemporaries, of course, did not use the term "Industrial Revolution," although its later participants and observers knew they had lived through something remarkable. They described it in the same terms used here, as a cotton manufacturing industry that developed in the north of Britain in the last decades of the eighteenth century, culminating in the period in which they were living – the 1830s and 1840s. These men include William Radcliffe and Robert Hyde Greg, Karl Marx and Edward Baines, all of whom we shall meet in Chapter 5, as our story comes to an end. By then, the events described here had begun to take shape as an historical incident with clear lessons to impart – a mythical history, devised to support a particular ideology and political goal – about innovation, and genius, and the role of technology in causing social change. Local industrialists celebrated their own acumen and won their political voice.

---

[3] David Landes, *Unbound Prometheus*, 1; Inikori, *Africans and the Industrial Revolution*, 156–65.
[4] P. J. Cain and A. G. Hopkins, *British Imperialism: 1688–2015* (Abingdon and New York: Routledge, 2016), 82.

Two generations later, in 1884, Arnold Toynbee generally receives credit for first naming the Industrial Revolution.[5]

By the time the Industrial Revolution received its name, the mythical version of events had already taken hold, commemorated in monuments and celebrations of heroic inventors, to whom were assigned the individual creation of ingenious machines that changed the world. Historians would spend the next hundred years trying to correct the schoolboy simplification that saw only a "wave of gadgets" that had "swept over Britain" to create that nation's nineteenth-century economic growth and imperial power.[6] Biographies of engineers circulated, as they do today, celebrating the heroic inventors of the tech industry. Victorian communities even erected statues to their favorites, to whom they credited their wealth. The phrase "Industrial Revolution," coined in 1884, simply expressed an invention myth, abridged from real events, that already resonated.

Technology in the Industrial Revolution intends to have it both ways. It treats the Industrial Revolution as real and revolutionary, but emerging from and encompassing long-term gradual shifts. It may seem contradictory to both accept the revolutionary nature of the period between the 1760s and 1840s and at the same time recognize its precursors and aftershocks, reverberating in long cycles around those decades, but such experiences abound in the history of technological change. Causation is complicated around machinery. Industrialization happened gradually, bit by bit, and then appeared suddenly, in the nineteenth century, and grew from there. One important reason industrialization happened successfully was the application of large-scale plantation agriculture to the cultivation of cotton fiber in the New World – but this happened during the Industrial Revolution, and accelerated as a result. Another cause of industrialization was the threat posed to an important national industry by the cotton textiles imported from India, and one effect was the colonization and imperial exploitation of British India. In these histories, mechanization stands between cause and effect, and partakes of both. The machinery at the heart of the story was only one part of a changing system and did not shift all the rest on its own. The belief that inventions caused industrialization was itself invented by industrialists who celebrated themselves and their industry in the nineteenth century. In that history, the reasons those machines worked then and there remains unexplained.

---

[5] Baines, *History of the Cotton Manufacture*; Karl Marx, *Das Kapital* (Hamburg: Verlag von Otto Meissner, 1867–1894); Arnold Toynbee, *Lectures on the Industrial Revolution in England: Popular Addresses, Notes, and Other Fragments* (London: Rivingtons, 1884).

[6] T. S. Ashton, *The Industrial Revolution, 1760–1830* (Oxford University Press, 1968), 48.

Debates and Terminology

In utilizing a longer timeline and larger framework, *Technology in the Industrial Revolution* sometimes employs contested terminology without criticizing the language, taking a side in the debates, or rationalizing the use of the term. For example, proto-industrialization is a troublesome word: it indicates an outcome (industrialization) and identifies steps that seem to lead there. It therefore implies that the outcome was pre-ordained, rather than contingent and uncertain at the start. This book uses the term "proto-industrialization" but tries to avoid assuming which ending is on the way. Instead it describes events emerging from local history and participating in global processes. Likewise, "industrialization" itself is defined by economists as the movement of resources from agriculture or extraction into making things "without much direct input of natural resources" – fabrication, in other words, rather than processing.[7] This book is interested in how that version of "making things" first originated. Examining the reasons behind the success or failure of particular production systems means seeing the way they fit into the world around them, even as that world was changing. Rather than assuming one technical system was better than another, the goal is to identify which elements in each system served what purposes.[8]

Guilds receive similar treatment: their interest for students of industrialization lies not in whether they were good or bad, beneficial or economically inefficient – points that professional historians debate. Instead, this book traces some of the steps by which guilds were transformed into new collective institutions, for capital and for labor, during industrialization. The reasons when and where and why the guilds declined matters less here than the permutation of some parts of guild structures into trade unions and business corporations. Likewise, this book attempts to explore a major transition point in the history of capitalism – from merchant to industrial capitalism – without arguing that one superseded the other, nor that mercantilism disappeared as industrialism emerged. This book defines "capitalism" as the investment of capital, put into operation in the hope of generating a return – capital risked on the assumption that wealth can be grown. While economists define capitalism as an economic system in which private entities own most property and make economic decisions free of centralized planning and government control, this book watches that ideal first emerge. And so it approaches

---

[7] John C. Black, *Dictionary of Economics*, 2nd ed. (Oxford and New York: Oxford University Press, 2002), s.vv. "industrialization" and "industrial sector."

[8] Regina Grafe, "Review of Epstein and Prak, *Guilds, Innovation, and the European Economy*", in *Journal of Interdisciplinary History* 40, no. 1 (Summer 2009): 78–82.

capitalism from the past that predates it, when public and private were harder to distinguish. The key is the investment placed at risk in order to grow.

Methods and Approaches: The Historiography of Technology

Wide historical research into the events known as "the Industrial Revolution" has created specialization. Each historical subfield has its own interests and questions. *Technology in the Industrial Revolution* both draws upon and speaks primarily to economic history and global history, and utilizes the work of labor and social historians and environmental historians as well. However, its guiding approaches and methods have been developed in the history of technology. Technology's historians use two basic approaches to understand how technology works: internalist and contextualist. Internalist investigation studies devices and methods without much reference to the world outside the technologies. Contextualist analysis does the opposite and incorporates the outside context – costs and prices; laws and institutions; cultural expectations about work, spending, and investment; and social norms around gender and age – into understanding the machinery. Internalist analysis tends toward technological determinism: machines appear and cause change, and each is better than the last. Contextualism slides toward social construction, in which the success of the technology is caused by events outside the machine. Over the past thirty years, historians of technology have moved toward reconciling the two approaches, in order to use the workings of machinery to understand both its causes and its effects.[9]

Combining contextual analysis with internalist understandings, Thomas P. Hughes's foundational *Networks of Power: Electrification in Western Society* fashioned what historians of technology call "systems theory" in explaining the development of large-scale technological systems like those that supply electricity in the western world. Rather than identify an inventor of the light bulb, he investigated the history of electrification in four different cities in Europe and the United States, and explained how the different sources of power supply, and the geography of customer bases, resulted in distinct technological systems for electrifying New York, Chicago, London, and Berlin. Hughes pointed out that machines develop and operate within systems, and politics and

---

[9] John M. Staudenmaier, S. J., *Technology's Storytellers: Reweaving the Human Fabric* (Cambridge, MA: MIT Press and the Society for the History of Technology, 1985), introduction and chapter 1.

economics play a role in how those took shape. In each city, the constellation of generators, transformers, transmitters, wires, outlets, and even light bulbs, in homes, businesses, and transport systems, had different designs. Over time, it became more and more difficult to change any particular part of the system. The parts fit together, and substantial changes to one segment would mean changing the rest – an increasingly unlikely undertaking, as the system matured and grew more complex. That is how, in Hughes's systems theory, the social construction of technological systems turns eventually into determinism as the system ages. Electricity – a system deeply influenced by politics, economics, and society – became harder to change as it influenced and shaped the world around it. Hughes' systems theory is perhaps most famous for its articulation of the "reverse salient," (borrowed from military history) in which bottlenecks to the development of a working system draw the attention of engineers and other systems-builders, so that the technology develops as a unified whole.[10]

In his later work, Hughes developed the concept of "stakeholders" in systems, including all the relevant social groups that influence a technological system. For example, building a bridge in a city can involve not just engineers but also politicians and planners, zoning boards and local businesses, and residents enrolled in advisory or protest groups. Construction contractors and materials suppliers also play a role in the design decisions. All these people and institutions and objects participate in how the system eventually is designed and built, and a successful engineer or project manager will have to manage all the stakeholders in order to advance the project. Such actors can be called heterogeneous engineers because their problem-solving uses more materials than those found in the physical world. This concept of stakeholders enlarges the analysis of a technological system beyond physical artifacts and the abstract power of politics, economics, and society. It includes human choices and organizations in the story of how technological systems took shape the way they did. The heterogeneous engineer is a useful concept for understanding the work performed by the renowned inventors of industrialization, whom we shall meet in Chapter 2.[11]

In expanding technical systems to include a wide range of actors and objects, the networks of their interactions, their economic

[10] Thomas P. Hughes, *Networks of Power: Electrification in Western Society, 1880–1930* (Baltimore and London: Johns Hopkins University Press, 1983).
[11] Wiebe E. Bijker, Thomas P. Hughes, and Trevor Pinch, eds., *The Social Construction of Technological Systems: New Directions in the Sociology and History of Technology* (Cambridge, MA: MIT Press, 1987; rev. ed. 2012).

interests, and the groups they participate in creating and maintaining, this book employs a rudimentary version of actor-network theory (ANT). Developed by Bruno Latour and other sociologists who work in the history of technology, ANT provides a way to connect people's actions and human agency to the forces that seem to operate outside mortal control. It replaces the concept of society, which sounds too settled, with the notion of "the social," which seems more contingent and ongoing than "society." In ANT, the social is an assemblage of people, institutions, and artifacts that are continuously reorganizing in ways that form the social realm. The theory avoids abstractions ("capitalism" or "patriarchy") and instead examines the practical steps people take, the institutions they create or join, and the artifacts they employ and develop, in their constant process of assembling their worlds. While historians of capitalism emphasize the forces that shifted human behavior from subsistence into market orientations, historians of technology avoid abstract causes, arguing that these forces only act through specific deeds. Rather than employing "capitalism" as a word that explains human behaviour, modern history-of-technology methods – and ANT in particular – instead trace the way that people and their actions embody capitalist activities.[12]

Another way historians study technology is to explore how it is used, rather than focusing on innovations. Museum professionals, for example, are shifting away from simple displays of machinery to find ways to explore and exhibit people's experiences of technologies. These approaches include considerations of the maintenance that keeps systems working. Daily use and the maintenance of existing devices are primary experiences that people have with technology. David Edgerton has criticized his fellow historians of technology for focusing too much on innovations, on "the early history of selected technologies which later came into widespread use." *Technology in the Industrial Revolution* pleads guilty to this charge. It also stems from an equally old-fashioned history-of-technology approach that some might call myth busting – undermining invention myths by identifying their contextual causes, adding complexity and nuance to the story of how technology was made to work. In other words, this book pursues what Edgerton describes as technology in history – a marriage of devices with their contexts. It argues that novelty draws upon elements that already exist, even as those change to get new things going. Today's

[12] Bruno Latour, *Reassembling the Social: An Introduction to Actor-Network Theory* (Oxford: Oxford University Press, 2005).

western, industrialized, capitalist societies rely on older understandings and institutions, including patriarchy, guilds and households, and mercantilist beliefs and policies hidden within an ideology of free markets.[13]

Earlier treatments by historians of technology have provided to this study a wealth of internalist knowledge. Contextualists, on the other hand, have often deployed the Industrial Revolution as an inflection point in the grand historical narratives of Big History, which usually point toward either progress or decline. In this role, the Industrial Revolution has become a black box – an unexplored technology, with known inputs and outputs but little examination of what happened in between. In these cases, mass production emerges from the Industrial Revolution, which therefore works to periodize world history to account for European economic and geographic expansion over the next century.[14] These approaches have fallen out of fashion at least partly because they either valorized or demonized what historians once called Western Civilization. Since scholars have abandoned these progress- or decline-oriented approaches, they have often abandoned the classic case of industrialization, the mechanization of British textile production in the late eighteenth century. Brewing, shoe-making, glass production, iron-working, and steam power have all become textbook cases. Meanwhile, scholarship on alternative cases has decentralized the British textile story from the history of industrialization. Daryl Hafter's and Leonard Rosenband's masterful works on French paper-making and textiles explore such options, and short sketches of their cases (and a few others) can be found in the Appendix of Alternative Examples.[15]

Economic History: Nature vs. Culture

Because industrialization is associated with mass production, economic growth, and European rise to world domination in the nineteenth century, economic historians have long sought explanations for its causes. A generation of arguments about rates of change can be

---

[13] David Edgerton, "Innovation, Technology, or History: What Is the Historiography of Technology About," *Technology and Culture* 51, no. 3 (July 2010), 687.

[14] Richard L Hills, *Power in the Industrial Revolution* (Manchester: Manchester University Press, 1970); Lewis Mumford, *Technics and Civilization* (New York: Harcourt, Brace and Co., 1934).

[15] Thomas J. Misa, *Leonardo to the Internet: Technology and Culture from the Renaissance to the Present* (Baltimore and New York: Johns Hopkins University Press, 2004).

found clearly summarized by Maxine Berg and Patricia Hudson in "Rehabilitating the Industrial Revolution."[16] For those who accept even a qualified premise of an Industrial Revolution, subjects of investigation have included finance and capital accumulation, the causes and effects of mechanization, and the rise and spread of the factory mode of production. The big question for workers in this field, has generally been: Why Britain? Why did the Industrial Revolution happen then and there? Answers have often focused on either economic or cultural factors. Within more cultural causes, institutions (property rights, patent law, and literacy levels, for example) contributed to industrialization, including Parliament's protectionist support for industry.[17] Enlightenment ideals, experimentation, and increasing popular access to formal knowledge also appear crucial to the technical innovations of the late eighteenth and early nineteenth centuries. Joel Mokyr has also identified the importance of micro-inventions – the small adaptations and devices that so often made the big-name inventions work.[18]

Explanations that rely more heavily on the natural advantages of economic factors (endowments of land, labor, and capital) include the way that Europe's indented coastline created competitive nation-states. Other nature-based explanations focus on the availability of coal for steam engines. E. A. Wrigley, for example, sees the shift to inorganic energy sources as the crucial breaker of economic constraints. Coal comprised half the total energy consumed in England as early as 1700, and a century later reached 75 percent, and then 90 percent by 1850. This expansion in the use of coal does indicate its increasing importance. For most of that time, though, most of the coal was for cooking meals and heating homes. Up to the end of the 1700s, more than half of Britain's coal was used domestically. In the next century, as the industrial form of production matured and spread, coal and steam engines helped that expansion. In the early nineteenth century, domestic consumers were using only about a third of the total coal, and then steam power "became massive in the economy" after 1825. We shall examine the periodization of

---

[16] Maxine Berg and Pat Hudson, "Rehabilitating the Industrial Revolution," *Economic History Review*, New Series, Vol. 45, no. 1 (Feb. 1992): 24–50.

[17] Patrick O'Brien, Trevor Griffiths, and Philip Hunt, "Political Components of the Industrial Revolution: Parliament and the English Cotton Textile Industry, 1660–1774," *Economic History Review* 44, no. 3 (Aug. 1991): 395–423.

[18] Ralf Meisenzahl and Joel Mokyr, "The Rate and Direction of Invention in the British Industrial Revolution: Incentives and Institutions," NBER Working Paper no. 16993 (Apr. 2011), www.nber.org/papers/w16993, accessed 6 February 2016.

power-source selection more deeply in upcoming chapters, alongside men who chose when to use which type of power.[19]

More recently, economic historians are debating whether high wages inspired the Industrial Revolution. Robert C. Allen explains the economic "incentive to invent coal-powered, mechanized technologies" with data showing high wages in England, especially relative to the low cost of coal. This update of the Habakkuk thesis makes the new technology a matter of cost-saving.[20] Certainly lowering the costs of producing textiles was one important effect of industrialization, but that does not mean it was the goal that people hoped to achieve. Moreover, machines were already developed and in use when coal-burning steam engines pushed or permitted them to increase production to mass levels. Effects are not the same as causes, and *Technology in the Industrial Revolution* is interested in the industrialization that occurred before steam power entered widely into producing goods. Although his claims apply primarily to the nineteenth century, nonetheless Allen does important work by shifting his subdiscipline's focus from invention to adoption. Critiques of Allen's high-wage explanations usually focus on his data and calculations, and find industrialization occurring in mostly the low-wage regions of Britain. Jane Humphries and Jacob Weisdorf, for example, have found lower wages than Allen did. Additionally, Allen used wages of builders to explain textile mechanization, but builders and spinsters seem unlikely to compete for one another's jobs. It is doubtful that these data will answer the question to which they have been applied. Debates continue.[21]

The question of wages indicates the difficulties faced by using economic-history approaches to understand technological change. The discipline's positivism directs practitioners to limit the factors under consideration until their relative contributions can be tested and measured. In contrast, historians of technology tend to add variables to the process of technological change, rather than reduce them to factors such as wages and coal prices, or property rights, or patenting costs. So *Technology and the Industrial Revolution* instead looks at a range of schemes, the choices

---

[19] G. N. von Tunzelmann, *Steam Power and British Industrialization to 1860* (Oxford: Clarendon Press, 1978), quotation at 4; E. A. Wrigley, *Energy and the English Industrial Revolution* (Cambridge and New York: Cambridge University Press, 2010), 38.

[20] Robert C. Allen, *The British Industrial Revolution in Global Perspective* (Cambridge and New York: Cambridge University Press, 2009), 33–34; H. J. Habakkuk, *American and British Technology in the Nineteenth Century* (Cambridge and London: Cambridge University Press, 1962).

[21] Morgan Kelly, Joel Mokyr, and Cormac Ó Gráda, "Roots of the Industrial Revolution," UCD Centre for Economic Research Working Paper Series, WP2015/24, (Oct. 2015); Jane Humphries and Jacob Weisdorf, "The Wages of Women in England, 1260–1850," *Journal of Economic History* 75, no. 2 (June 2015): 405–47.

individuals made, and the institutions they created, as well as the objects they employed. Instead of "Why Britain?," this book shifts the question slightly to "Why did those machines work there and then, in that specific time and place?" Contingent moments, some of them detailed in the text of this book, contributed decisively to the result. Choices can be reduced to costs, but costs were not the only considerations, as will become apparent as different machines were deployed over time. Sometimes industrialists were willing to pay more to buy and operate machinery in order to replace cheap but rebellious workers – as we shall see in Chapters 4 and 5. In addition, costs change when technology does. For this reason, prices are but a partial explanation, at best, for technological change.

Economic historians who emphasize culture, literacy, and experimentation speak as well to the history of science, as the scholars in that field are primarily interested in how knowledge is made and used. These approaches have provided a wealth of information about the know-how circulating during the Industrial Revolution, and manufacturers' exposure to it. *Technology in the Industrial Revolution* instead uses the approaches of historians of technology, and foregrounds what manufacturers did rather than what they knew.[22] After all, the word "scientist" was first coined in 1834. Instead of reading the occupation back into the period before it existed, this book instead relies on the research by scholars, including Pamela Long and Pamela Smith, who study early modern science, art, and knowledge, through a longue durée of global trade in crafts and technique. They study the sources of knowledge and the artisanal background of the engineers who eventually built textile machinery, waterwheels, and steam engines. Their findings are a better fit for the longer timeline explored here.[23]

National and Global History

Mechanization of British textile production between 1760s and 1840s: this classic case of Industrial Revolution took place primarily in one branch of business in a few counties in the north of England. In Britain,

---

[22] Joel Mokyr, *The Enlightened Economy: An Economic History of Britain 1700–1850* (New Haven: Yale University Press, 2012); Margaret C. Jacob, *The First Knowledge Economy: Human Capital and the European Economy, 1750–1850* (Cambridge and New York: Cambridge University Press, 2014).

[23] Pamela O. Long, *Openness, Secrecy, and Authorship: Technical Arts and the Culture of Knowledge from Antiquity to the Renaissance* (Baltimore: Johns Hopkins University Press, 2001); Cormac Ó Gráda, "Did Science Cause the Industrial Revolution?," University of Warwick, Working Paper Series #205, www2.warwick.ac.uk/fac/soc/economics/research/centres/cage/manage/publications/205-2014_o_grada.pdf, accessed 11 May 2016.

therefore, it is usually an episode of national history. Interpretations of these events have thus been influenced by the national mood in which each historian wrote, including the triumphant Victorian version being constructed at the end of this book. A useful guide through this scholarship can be found in the 1984 essay by British historian David Cannadine. He linked British historians' interpretations of the Industrial Revolution to the periods of national history in which they lived. For example, Cannadine argued that the inflation and unemployment of the 1970s, the unwinding industrial system, shaped the pessimism of those who argued that industrialization had never been revolutionary to begin with. Talented generations of scholars have teased out a wealth of details and considerations, and have also produced comparisons between Britain and other places – especially Continental Europe, where some nation-states shared many of Britain's conditions, and yet did not make the shift to mass production quite as early or as dramatically as Britain had done. Other nations' changes to production technology were only rarely formulated into a mythical version such as the Industrial Revolution.[24]

The Industrial Revolution was not only local and national history, it was a part of larger global processes (as is true of all history, of course). Groundbreaking early scholarship in global history linked agriculture, mining, and slavery in the New World to industrialization and European imperial power. More recent historians have shifted the global picture from the Atlantic world to the Indian Ocean, and asked why Europe grew rich while Asia, once the world leader both in wealth and in textile production, lost its earlier preeminence. Giorgio Riello's global history of *Cotton* argues that European industrialization was epiphenomenal to shifting global commercial patterns. He includes cloth scraps from the eighteenth and nineteenth centuries as his evidence for trade circulation, raw materials, and machine processes. Prasannan Parthasarathi has jettisoned the notion that superior European institutions explain the divergence from Asia and instead explores similarities among European and Asian institutions, from property rights and commercial connections to demographic trends and scientific communities. Drawing on economic theories that admit many possible outcomes from similar factor endowments, Parthasarathi's symmetrical analysis shows that India was not an outlier. In studying these similarities, it becomes clear that England's industrialization was interwoven into world trade

---

[24] David Cannadine, "The Present and the Past in the English Industrial Revolution, 1880–1980," *Past and Present* 103, no. 1 (May 1984).

and the mercantilist government policies that turned commerce into empire.[25]

More textile-based scholarship can often be found in the books produced by the Pasold Studies in Textile History, which often draw on scholarship in design and fashion history. What people wore, how they chose and maintained their clothes and soft furnishings, drives the history of textile industrialization. Beverly Lemire's indispensable work on cotton and consumers has made plain the extent that consumption came first and stimulated industrialization. Neither superior supplies of environmental nor institutional factors explain as much to Lemire as the changing demand for light, bright, comfortable cloths. As John Styles points out, however, the English "Calico Craze" identified by Lemire may have been exaggerated by domestic woolen and worsted merchants seeking Parliamentary protection for the domestic market, within which industrialization thrived. Understanding merchant efforts to meet this domestic demand, and to compete overseas, results in a longer story that includes new mixtures, fresh products, and eventually even tasks embodied in machinery. Scholars of consumption have also demonstrated the extent to which cloth making had specialized into crafts performed outside the home by the eighteenth century, setting the stage for later mechanization.[26]

In arguing that industrialization grew within an existing industry, *Technology and the Industrial Revolution* draws on the work of scholars who have researched fibers other than cotton. From Yorkshire's woolen and worsted industries, Pat Hudson's *Genesis of Industrial Capital* provides the basis for understanding not just the origins of capital for later industrialization but also the family and business structures that framed and shaped mechanization. John Smail's studies of Yorkshire merchants emphasize both consumer demand and the importance of individual decisions, made in market conditions. These works revise the association of industrialization with mechanization by pointing out the many important transformations that did not include machinery. In addition, seeing cotton as just one of many fibers places industrialization into a global and regional competition of production, demand, and commercial relationships. So too does the fundamental work of Maxine Berg, who in *The Age*

---

[25] Parthasarathi, *Why Europe Grew Rich and Asia Did Not*; Giorgio Riello, *Cotton: The Fabric That Made the Modern World* (Cambridge and New York: Cambridge University Press, 2013).

[26] Beverly Lemire, *Fashion's Favorite: the Cotton Trade and the Consumer in Britain, 1660–1800* (Oxford and New York: Oxford University Press, 1991); John Styles, "Indian Cottons and European Fashion, 1400–1800," in Glenn Adamson, Giorgio Riello, and Sarah Teasley, eds. *Global Design History* (London: Routledge, 2011), 37–46.

of Manufactures reminded scholars and students of the multiple sectors – not just textiles – in which consumer demand stimulated mechanization and new forms of organization.[27]

Subdisciplinary Differences

Social and labor historians have often made the Industrial Revolution one of their central concerns, and E. P. Thompson's *Making of the English Working Class* is foundational work on the topic. Thompson built on nineteenth-century analysis by Karl Marx and Friedrich Engels to find industrialization the origin point of modern class identities. Historian Malcolm Chase has neatly summarized this scholarship in his history leading up to *1820.*[28] Chase seeks primarily to explain nineteenth-century British political reform movements and the rise of ideals of representative government from their working-class origins. Many labor historians focus on labor organizing and resistance, but the work of Carolyn Steedman represents an important exception. Steedman's study of domestic servants points out understudied areas of working-class formation, neglected perhaps because the workers were women, lived where they worked, and produced no physical objects from their *Labours Lost.* Another source-base for the labor history of early industrialization appears in worker autobiographies, but these sources have led to divergent interpretations of workers' experience. Jane Humphries' systematic investigation of the memoirs of former child laborers found disadvantages both pulling children into work, and deriving from their early employment – some were even crippled by their early labor. However, Emma Griffin has used similar first person accounts to find a more positive reading of the opportunities people encountered in the new labor systems.[29]

Meanwhile, environmental scholars, both scientists and humanists, also view the Industrial Revolution as the vital step on the path to modern dependence on fossil fuels. They have named the result the Anthropocene

---

[27] Pat Hudson, *The Genesis of Industrial Capital: A Study of the West Riding Wool Textile Industry, c. 1750–1850* (Cambridge: Cambridge University Press, 1986); John Smail, *The Origins of Middle-Class Culture: Halifax, Yorkshire, 1660–1780* (Ithaca and London: Cornell University Press, 1994); Maxine Berg, *The Age of Manufactures: Industry, Innovation, and Work in Britain, 1700–1820* (Oxford: Basil Blackwell, in association with Fontana, 1985).

[28] E. P. Thompson, *The Making of the English Working Class* (London: Victor Gollancz, 1964); Malcolm Chase, *1820: Disorder and Stability in the United Kingdom* (Manchester and New York: Manchester University Press, 2013).

[29] Carolyn Steedman, *Labours Lost: Domestic Service and the Making of Modern England* (Cambridge and New York: Cambridge University Press, 2009); Jane Humphries, *Childhood and Child Labour in the British Industrial Revolution* (Cambridge and New York: Cambridge University Press, 2010); Emma Griffin, *Liberty's Dawn: A People's History of the Industrial Revolution* (New Haven: Yale University Press, 2013).

(or even the Capitalocene) – a new geological period defined in terms of human impact on the earth. Within this field, the historian Edmund Russell notably describes industrialization in terms of energy flows, and argues also that the contribution from cotton varietals that suited the new machinery cannot be ignored. Andreas Malm has written a riveting account of the transformation from organic to inorganic power sources in *Fossil Capital*. On its own, however, coal was inert. It had existed for most of human history, but it needed technology to transform it into an efficient source of power for mass production. For this reason, *Technology and the Industrial Revolution* will emphasize industrialization before coal and steam engines achieved the dominant role they would claim in Victorian Britain. As with cotton from the American South, coal did not cause the Industrial Revolution. Both came too late to cause mechaniza-tion, though both were crucial to the widespread adoption and success of those devices. As Malm recognizes, the timing and causation of the switch to using fossil fuels was complicated and indirect. It was only *during* the process we now call the Industrial Revolution that coal became an energy source for mass production – and the process was contingent on local events.[30]

Organization

Technology in the Industrial Revolution begins by exploring the world from which industrialization took shape. Its first chapter sketches textile history on the British Isles, before examining their accelerating interactions with the rest of the world. The chapter is named "Sugar and Spice" for the early-modern predecessors of industrial capitalism: the plantation mode of production that developed in large-scale sugar cultivation and that later would supply cotton to the mills of the Industrial Revolution, and the trade in spices that linked Europe to Asia and sparked a new phase of international competition in textile production and markets before 1760. Domestic production, proto-industrialization, and guilds sketch Britain's textile sector before industrialization. Cloth imported by the East India Company shocked and disrupted this industry, which turned to govern-ment protection, within which its merchants experimented and thrived.

Chapter 2 shifts focus to changing machinery. It both provides and undermines a traditional, linear, internalist version of the technical changes that industrialization ensconced – a succession of machines for

---

[30] Edmund Russell, *Evolutionary History: Uniting History and Biology to Understand Life on Earth* (New York: Cambridge University Press, 2011); Andreas Malm, *Fossil Capital: The Rise of Steam Power and the Roots of Global Warming* (London and New York: Verso, 2016).

spinning and how they worked. Associating these machines with the men who invented and adopted them helps us understand the contexts within which they lived and operated. Their worlds indicate what external inputs went into making famous machines work. From the pauper children apprenticed to work in spinning mills to the American plantations that adopted cotton cultivation using slave labor, new machinery worked by utilizing existing sources of supplies, even as they were changing. The decisive element in mechanizing the cotton industry was Richard Arkwright's successful Parliamentary maneuvering, which carved out an exception to the Calico Acts that made cotton spinning profitable for those who used his system.

Chapter 3 then extends the story beyond individual machines to the development of the entire industry. It focuses on Manchester, the town in Lancashire County most closely associated with the Industrial Revolution. It was known as Cottonopolis, though all the cotton it ever sold was grown somewhere else. Both celebrated and reviled as the site of the world's first factory district, Manchester's local history – both its canal infrastructure and the technical choices its people made – contributes to the story of which machines worked and became part of the Industrial Revolution narrative. Cottonopolis links Manchester to slave factories in Africa and plantations in North America, as well as to the cotton industry of India, to demonstrate the reverberations between technological change and its widening contexts.

Chapter 4, "Power and the People," explores the changing power sources for running textile machinery. Comparing emerging industrial systems demonstrates the contingent reasons for adopting various parts of the industrial idea, including steam power. In Britain's textile industry, steam engines (and the capital invested in them) were sometimes the cause of worker unrest, and sometimes the result. The Luddites who broke textile machinery in industrializing districts were responding in traditional ways to changing employment patterns. Their legendary resistance to technology is actually part of a longer and more complicated story. Their actions are part of the process of technological change, as well as revolt against it – they were some of the reasons that one technology's advantages came to outweigh others. On the other hand, Manchester's capitalists incorporated steam-driven Iron Men in their spinning mills as a response to worker unrest, to deliberately break the strike of workers seeking higher pay. As working-class identity merged into political rebellion, so did capitalists marshal government support for their purposes. The support that authorities offered to capital, rather than labor, became apparent in the Peterloo massacre of 1819.

Finally, in Chapter 5, the spread of steam-powered, mechanized weaving in the nineteenth century illustrates the integration of production into a "Vertical Mill" that reorganized global trade patterns and separated consumption from production for increasing numbers of people worldwide. By the 1840s, when the book ends, industrialization had incorporated the development of new class structures. Both workers and industrialists used their social class – their relation to the means of production – as the basis for their political power. The concept of invention was itself invented, a buttress to industry's ideals that achieved specific political goals when Parliament repealed the Corn Laws in 1846. This accomplishment enshrined an ideology of free trade and a mythology of laissez-faire that accurately described neither the past from which industrialization had sprung nor the imperial nation then coming into being. Instead, the myths of invention served the goals of an industry founded from existing elements, including the government protection and mercantilism that it tailored to fit new purposes as it grew.

The Industrial Revolution was a revolution in many spheres. Even those scholars who admit its existence still ask: was it a demographic, economic, technological, cultural, social, or political revolution? The answer is: yes. It was all of these. Technological systems incorporate all these domains. The distinctions among them exist for scholars and students, but historical actors experienced them all at once. This book attempts to look at least briefly into all these realms, to see how the technology developed in all the contexts from which it drew.

Suggested Readings

Berg, Maxine, and Pat Hudson. "Rehabilitating the Industrial Revolution." *Economic History Review*, New Series, 45, no. 1 (February 1992): 24–50.

Frank, Andre Gunder. "A Plea for World Systems History." *Journal of World History* 2, no. 1 (Spring 1991): 1–28.

Hahn, Barbara. "The Social in the Machine: How and Why the History of Technology Looks Beyond the Object." *Perspectives on History: The Newsmagazine of the American Historical Association* (March 2014): 30–31.

Horn, Jeffrey, Leonard N. Rosenband, and Merritt Roe Smith, eds. *Reconceptualizing the Industrial Revolution.* Cambridge, MA: MIT Press, 2010.

Mokyr, Joel. *The Enlightened Economy: An Economic History of Britain 1700–1850.* New Haven: Yale University Press, 2012.

Parthasarathi, Prasannan. *Why Europe Grew Rich and Asia Did Not: Global Economic Divergence, 1600–1850.* Cambridge and New York: Cambridge University Press, 2011.

Riello, Giorgio. *Cotton: The Fabric That Made the Modern World.* Cambridge and New York: Cambridge University Press, 2013.

Russell, Andrew L., and Lee Vinsel. "After Innovation, Turn to Maintenance." *Technology and Culture* 59, no. 1 (January 2018): 1–25.

Vries, Jan de. *The Industrious Revolution: Consumer Behavior and the Household Economy, 1650 to the Present.* Cambridge and New York: Cambridge University Press, 2008.

1 Sugar and Spice

The word "innovation" indicates novelty. In order to make their way, however, new things draw on those that already exist. So it was with the grand historical systems and processes called capitalism and industrialization; so it is with the humbler systems and processes that people use to clothe their bodies and furnish their homes. Long before the historical record focuses on cloth, most everybody made it. Ordinary families made the clothing and towels and bedding needed by their households, while intricate networks traded luxury goods over long distances. We know some things about the goods that entered the market, less about the pieces made at home for daily use. One historian of British medieval cloth-making declared, "activities need something more than universal importance if they are to take their due place in history," but studying the Industrial Revolution makes fabric a crucial element of global history.[1] This chapter sketches a rough picture of how people made and acquired textiles in some preindustrial societies, and then how cloth wove a few of those worlds together. It traces the preindustrial history of capitalism, in which merchants invested capital in commerce, and governments fortified and regulated and taxed their activities. Well before the Industrial Revolution of the late eighteenth century, goods traveled the globe. Networks of people and their systems for moving merchandise provided the basis for the technological shifts associated with industrialization. From world shipping patterns to individual households and family structures, the emergence of capitalism and nation-states before industrialization set the stage for the events that would transform them both.

Demography

Slow changes in population frame the industrialization story. People are both workers and consumers, and long-term increases in both labor and

[1] A. R. Bridbury, *Medieval English Clothmaking: An Economic Survey* (London: Heinemann Educational Books and the Pasold Research Fund, 1982), vii.

demand played pivotal roles in the mechanization of textile production that took place in the eighteenth and nineteenth centuries. So let us begin in the late Middle Ages, when – between 1250 and 1600 – increasing English population increased agricultural output. Commercial activity also grew for nearly three centuries after 1086, counted by the number of markets and fairs in operation. Markets received charters from the Crown and in exchange paid the government, providing revenue. The increasing business, though, received a severe check in the fourteenth century. English medieval population peaked in 1300 at about six million people, before plague between 1347 and 1349 killed more than a third of all Europe's inhabitants. The plague continued to strike periodically, then occasionally, until the 1670s. By that time, though, in the two centuries after 1600, English advances in agricultural technology and productivity were reaching revolutionary levels.[2]

England's prolific agriculture fed fewer people than it could have done. In the 1650s, women in England married late or not at all (an extreme case of the so-called "European marriage pattern"). As a result, they had few children compared to the rest of the world, and fewer than half the biologically possible number. Limited population increased both wages and standards of living. While life expectancy averaged about thirty-seven years of age until 1800, most deaths occurred in birth and childhood – living beyond age twenty could mean living into your 70s or 80s. After 1730, however, birth rates sped up. More women were marrying at younger ages. Men with skilled trades and good prospects also married quite young, while illegitimate births also increased. Fewer women died in childbirth. Population grew into the 1730s and mortality declined as well: 1742 was the last year in England when deaths outnumbered births. The larger number of babies born in the 1730s themselves reproduced at the new higher rates in the 1750s, and there was more food available to feed these generations due to slight increases in temperature and productivity, and that food was cheaper. Extra food gave workers more capacity, and industriousness generated surplus resources. More people with more wealth bought more things. They made economic decisions that allocated resources toward consumer goods like better bedding, furniture, and kitchenware, as well as nicer clothes and pocket watches.[3]

[2] Gregory Clark, "The Long March of History: Farm Wages, Population, and Economic Growth, England 1209–1869," *Economic History Review* 60, no. 1 (Feb. 2007): 97–135; John Langdon and James Masschaele, "Commercial Activity and Population Growth in Medieval England," *Past and Present* 190, no. 1 (Feb. 2006), 43–45.

[3] Gregory Clark, *A Farewell to Alms: A Brief Economic History of the World* (Princeton and Oxford: Princeton University Press, 2007); Jan de Vries, *The Industrious Revolution: Consumer Behavior and the Household Economy, 1650 to the Present* (Cambridge and New York: Cambridge University Press, 2008).

More people with more food: eighteenth-century England was already escaping the trap predicted by Thomas Malthus in which the benefits of increasing productivity would be swallowed by growing populations. Despite annual variations, the expansion continued: from 4.9 million people in 1680, the nation's population more than doubled to 11.5 million in 1820. More better-fed people meant more fit workers to drive economic growth, and their numbers also testify to an expanding market for goods – both of which mattered for mechanizing production. The people were different from one another, however – in social status, access to funds, in their style of life. These differences became another resource that contributed to industrialization, even as those social structures changed.[4]

Land and Labor

Precedence and position ordered medieval English society. Lords had manors and controlled land and other resources, including labor. Worked by tenants (sometimes called serfs, villeins, or peasants), the land provided income to its owners, and agriculture dominated the economy. In exchange for various privileges of title and income, the lords, peers, and aristocrats provided material and military support to the Crown. The church and its monasteries likewise controlled land and supported surrounding communities. Everyone served someone and with superior status came inferiors who served their betters. The resulting web of mutual obligation was not as romantic as the tales of King Arthur and the Knights of the Round Table but neither was it a completely static system. A merchant might buy a manor that provided income from land, while other men were made aristocrats as a reward for military service. Some children raised in the countryside became apprentices and then guilded craftsmen, the artisans who built and decorated the great cathedrals (for example) in market towns. Even small villages supported specialized work, as occupational names indicate: Middle English had more than 165 surnames related to cloth (Weaver, Webber, Webster, Fuller, Draper, Taylor . . .). Nonetheless, servants, tenant farmers, knights and pages, lords and masters, king and clergy: the order of society was clear,

---

[4] E. A. Wrigley, "The Growth of Population in Eighteenth-Century England: A Conundrum Resolved," *Past and Present* 98, no. 1 (Feb. 1983): 121–50; Emma Griffin, "A Conundrum Resolved? Rethinking Courtship, Marriage, and Population Growth in Eighteenth-Century England," *Past and Present* 215, no. 2 (May 2012): 125–64; John Smail, *The Origins of Middle-Class Culture: Halifax, Yorkshire, 1660–1780* (Ithaca and London: Cornell University Press, 1994), 11–12.

and population growth still threatened to eat any possible gains in productivity.[5]

Nearly all medieval English families farmed the land both for its owners and for their own subsistence. Most people neither needed nor received much more than they could make in their own families or trade at a nearby market. What they ate, they made. What they wore, they also mostly made. The individual tasks of making cloth belonged to particular members of the family. Male heads of households did the weaving, completing the tasks of cloth making. The looms on which they worked embodied and reflected self-sufficiency, as it was possible to make a loom at home out of wood. Operating that loom, though, required inputs from the whole family. Women and children and grandparents prepared the wool and also spun it into yarn, the individual strands of thread to be woven into cloth. Men sheared the sheep to produce the fiber that the rest of the family prepared and spun, and then wove their yarn into the cloth that the household used in clothing, bedding, and toweling. The patriarchal family structure drew upon and supported this production technology, as well as the larger social structure, with its distinctive rights and obligations.[6]

An old English proverb declares, "the sheep hath paid for all." The fleece of sheep, sheared off seasonally, was the fiber immemorial of British textile production. English wool – the raw unprocessed fleece – was highly desirable in medieval continental Europe. By 1400, the export from England of woolen cloth usually exceeded the export of raw wool.[7] Tasks that turned fleece into finished cloth included preparation, spinning, weaving or knitting, and finishing. Preparation involved carding the fiber or combing it out and softening fibers until they lay in one direction, and the soft fragile untwisted rope they made was known as the sliver (See Figure 1.4). Drawing that sliver out narrowed it into a roving, a tighter, slightly twisted fleecy web of ribbon now ready for spinning, which made yarn. Spinning drew out the fibers even further and also twisted them tight, to create both strength and continuity – a long yarn for weaving into cloth. Drop spindles had been used since antiquity, but spinning wheels made the action more continuous in the medieval era, and may have tripled each worker's productivity. The device represented so important a shift in productive

---

[5] James Masschaele, *Peasants, Merchants, and Markets: Inland Trade in Medieval England, 1150–1350* (New York: St. Martin's Press, 1997), 14; John S. Lee, *The Medieval Clothier* (Woodbridge: Boydell Press, 2018), 39.

[6] Pat Hudson, *The Genesis of Industrial Capital: A Study of the West Riding Wool Textile Industry, c. 1750–1850* (Cambridge: Cambridge University Press, 1986), 59–64; Maxine Berg, *The Age of Manufactures: Industry, Innovation, and Work in Britain, 1700–1820* (Oxford: Basil Blackwell, in association with Fontana, 1985), 215–17.

[7] Bridbury, *Medieval English Clothmaking*, vii–xi.

methods that historian John Styles has termed the centuries 1400–1800 the "Era of the Spinning Wheel."[8] The innovation of spinning by wheel later became part of the European cultural mythology. When the Brothers Grimm collected fairy tales in the nineteenth century, they included the medieval story of Sleeping Beauty. She pricked her finger on a spindle and fell under a spell. In practice, spindles were not sharp enough to break skin, let alone induce a hundred years' slumber, but the hypnotic spinning wheel could still sometimes carry meanings of mystery and danger, even to readers no longer familiar with the device.

Spun yarn was readied for weaving as either the warp or the weft. A warp is the basis of the fabric being made – the warp is an array of long yarns, all oriented in one direction, strung vertically before the weaver, in width as far as he could reach to weave, in length as long as the finished piece of cloth will be. Through this warp, the weaver wove the weft, uncoiling the warp yarns to his work, rolling the woven fabric onto a beam before him. Warp yarns had to be sturdy: the loom constantly moved them against the wood. Each one would be threaded through a heddle, which was a vertical span of reeds like organ pipes, with another one (or more) behind it (see heddles at E in Figure 1.1). Lifting a heddle would lift the warp yarns threaded through it, and the gap between the lifted threads, and those that stayed in place, was called the shed. Then a weft thread (wound onto a shuttle or a quill – at S in Figure 1.1) would be passed through the shed, and then pressed tight against the other woven weft yarns. The weaver would then lift a different heddle, through which other warp yarns were strung, and thereby lift those other warp yarns. The weaver then passed the weft back through the new shed. Each pass of the weft, each woven yarn, was known as a pick. Each pick wove a weft yarn over some warp yarns and under others. The next pick carried the weft under the warp yarns it had passed over before. This was weaving. In woolens, the resulting cloth was somewhat loose and web-like, so fulling was the task that readied it for its eventual use as clothing or covering.

Fulling (or felting) meant pounding or trampling the woven cloth in a wet slurry containing fuller's earth, a clay mined from seams across southern England. The water and the pounding felted the loose woven wool, which made it thick, sturdy, and weather-resistant. Fulling can be done by hand or foot, and the pounding broke up the fibers to reform into a tight felt, while the clay absorbed the oil

[8] John Styles, "Fashion, Textiles and the Origins of Industrial Revolution," *East Asian Journal of British History* 5 (Mar. 2016): 161.

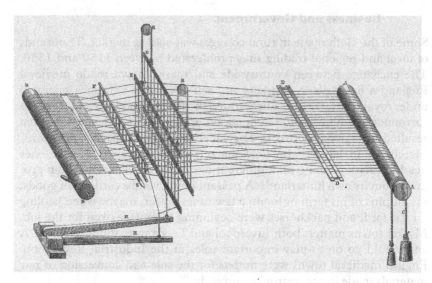

Figure 1.1 Weaving means pass a weft yarn through a warp. The weft, wound onto the shuttle S, was passed through the shed formed when a heddle at E lifted some of the warp yarns spread vertically before the weaver. Treadles at H controlled the motion of the heddles, and finished cloth was wound onto a beam at B. Courtesy of ZU_09/DigitalVision Vectors/Getty Images.

from the sheep's fleece as well as any dirt that had accumulated during production. For this reason, fulling also whitened the cloth. Urine, collected from the family or cattle or sometimes the neighbors, could do the same work as fuller's earth. Dyeing was usually performed on cloth already fulled. Woad made cloth blue while madder produced red and weld added yellow. Mordants facilitated and fixed the color. Madder and woad were imported from the Continent, but the rest of the chemicals could be found at home – specialists dealt in dyes, and dyeing was itself an expert's job. Whether colored or white (as dyed and undyed cloths were named), plain or rayed (striped) or mottled, wool cloth pieces were finished by techniques known as cropping and dressing, and sometimes other special treatments as well, to produce the desired appearance. Felted cloth was hung on tenterhooks and stretched on frames to dry and tighten up.[9]

[9] Michael Dickenson, "The West Riding Woollen and Worsted Industries, 1689–1770: An Analysis of Probate Inventories and Insurance Policies," (Ph.D. Diss.: University of Nottingham, 1974), 41–45, 61–72, 138–39.

Business and Government

Some of the cloth made in rural cottages was sold at market. Thousands of local and regional trading sites proliferated between 1150 and 1350. The exchange between countryside and market town made medieval England a lively place. Markets and self-governing boroughs operated under royal charters, and concentrated the flow of goods to and from the surrounding countryside. Competition among chartered markets often resulted in regional networks of smaller and larger villages and towns. Merchants in larger, wholesale markets could travel to smaller ones nearby, each operating on a different day of the week, each selling produce from its own hinterlands. A peasant might vend a cartload of goods, the surplus of his farm or loom, a few times a year, maybe when hauling for the lord, and packhorses were beginning to replace oxen for the job. Market towns matter: both Liverpool and Leeds were founded in 1207, and would go on to play important roles in the Industrial Revolution. English medieval towns were outlets for the sale and conversion of raw materials made in the nearby countryside.[10]

Food and victuals, hides and leather, and wool and textiles were the principal trades of the towns. Urban craftsmen of medieval England were often organized into guilds, in which groups of artisans or merchants gathered together and paid for monopoly rights to practice their particular trade in their town. A weavers' guild in York in 1164, for example, paid £10 into the Royal Exchequer per year for a charter that said that only its members could legally make certain types of cloth within the city. Guilds limited their membership, regulated the training and assessment of new members, and controlled the quality and prices of the goods their members made and sold. Their hierarchies were similar to those of families in patriarchal households. Apprentices, like children in the household, lived and worked in the homes and workshops of masters, learning the trade in an indentured training period that lasted usually seven years. At the end of their training, they became journeymen and could tramp around among towns and workshops, gaining more experience before setting up shop as masters themselves. Urban guild members shared local sources of raw materials and prohibited showing these connections to outsiders; their regulations prevented competition among guild members, and at the same time gave them rights and privileges over outsiders.[11]

---

[10] Masschaele, *Peasants, Merchants, and Markets*, 2–17, 73–77, 110, 189–211.
[11] Masschaele, *Peasants, Merchants, and Markets*, 17, 131, 138; Herbert Heaton, *The Yorkshire Woollen and Worsted Industries, from the Earliest Times up to the Industrial Revolution*, 2nd ed. (Oxford: Clarendon Press, 1965), 27–47.

Guilds and towns often shared structural similarities as corporations: legal entities comprised of multiple members and embodying their rights, prerogatives, and obligations. Incorporation was rare, a gift of the Crown. It granted privileges to guilds, towns, and universities in exchange for revenue, and guilds helped their towns maintain orderly business. Some boroughs or guilds restricted market activities to a guildhall or a trade hall, a space dedicated to exchange, sometimes even in a specific thing: Blackwell Hall in London, for example, was the place for merchants to sell the cloth bought from markets around the country. Even general markets were privileges that set limitations and regulations on business – "forestalling," or selling goods before they reached the market stalls, was a common crime. Guild members were sometimes called freemen, signifying their rights to trade in the town. They could even graze their sheep on the town commons, indicating how closely related were the two institutions. But town and guild interests only sometimes aligned. Urban corporations could control the business of the town – could make sure food was available at a fair price, for example – by either enforcing or ignoring guild rules regarding outside trade.[12] Guilds can be seen as early labor organizations and mutual-aid organizations, but their corporate structures and organizational hierarchies meant that although they encompassed the whole trade, they represented masters more than journeymen and apprentices.

Local and Global

The reference to London merchants hints at the international trade in textiles. We know that business was established before 1400 because that is when exports of finished cloth exceeded those of raw English wool (though, of course, proportions shifted back and forth in succeeding years with the tides of commerce). English fabric, sold in individually woven pieces, had customers and competitors across the European continent, in Holland and Hamburg, Flanders and Vienna and the north of Italy. In the second half of the century, London merchants increasingly controlled the finance and marketing networks that carried cloth overseas, and city regulations limited their competition. They organized most of the over-seas trade through commercial guilds, including the Merchant Adventurers of London, and the Worshipful Company of Drapers (who

---

[12] Ron Harris, *Industrializing English Law: Entrepreneurship and Business Organization, 1720–1844* (Cambridge and New York: Cambridge University Press, 2010), 17, 23; Michael John Walker, "The Extent of the Guild Control of Trades in England, c. 1660–1820; A Study Based on a Sample of Provincial Towns and London Companies," (Ph.D. Diss., University of Cambridge, 1985), 61, 94–96, 101–07.

sold finished cloth). London's cloth merchants could provide goods on credit to their European buyers, and in this way they furnished a financial network linked to cloth production.[13]

As English wool found markets overseas, so too did imports arrive, especially luxury fibers and fabrics. Importing exotic cloth and embroideries for prestige purposes (in church, for example) pre-dated even the Norman invasion of 1066. Medieval elites had access to material goods and design elements borrowed from distant places, including flowers and porcelain, spices and textiles. Fancier cloth was reserved for people of higher ranks – between 1337 and 1553, English sumptuary laws dictated who could wear what. Only the uppermost nobility could wear gold and silver cloth, for example. Such limits provided visual indications of social rank. Reserving decorative silks for high society also protected the domestic woolen industry against imports. Middle-ranked elites sometimes pushed the boundaries of the laws. They might line their clothes with silk to hide the violation, or find substitutes for forbidden materials. The popularity of violet garments among town dwellers and rural workers alike may have been a visual echo of the purple silks reserved for royalty. Increasing legal sumptuary specifications, and efforts to circumvent them, contributed to acceleration in changing fashions in the medieval and early modern world.[14]

Competition and specialization marked the medieval international textile trade: Italy's weavers made terrific silk, for example, and individual city-states were known for specific products. Part of the work of guilds was to maintain such reputations of type and quality. At the same time, however, one fabric could often be substituted for another, to serve a similar purpose, and textiles competed for consumers. The silk-producing cities of Italy made specialties but also challenged each other with novel goods and imitations, sparking a fashion cycle in which rapid replacement of clothes better suited buyers than sturdy goods made to last. The range of fabric types increased when the Crusades carried cotton yarn into Spain in the ninth and tenth centuries. However, manipulating the fiber was tricky for those unfamiliar with how to work it – Europeans could not reliably spin cotton into yarn strong enough to serve as warps.

---

[13] John H. Munro, "Medieval Woollens: Textiles, Textile Technology and Industrial Organisation, c. 800–1500", in D. T. Jenkins, ed., *The Cambridge History of Western Textiles*, Vol. 1 (Cambridge: Cambridge University Press, 2003).

[14] Hilary Doda, "'Saide Monstrous Hose': Compliance, Transgression and English Sumptuary Law to 1533," *Textile History* 45, no. 2 (2014), 173, 180–86; Gale R. Owen-Crocker, "Brides, Donors, Traders: Imports into Anglo-Saxon England," in Angela Ling Huang and Carsten Jahnke, eds., *Textiles and the Medieval Economy: Production, Trade, and Consumption of Textiles, 8th–16th Centuries* (Oxford and Philadelphia: Oxbow Books, 2015), 65–73.

So Europeans generally used cotton in blends, what they called mixtures: a cotton weft woven through a linen warp created fustian, while bombazine was cotton woven through silk. Medieval product innovations mixed cotton, linen, and wool into new fabrics, and these new cloth types both contributed to and drew from technological advances in production. Cotton was often a cheap addition to a more elegant fabric, and thus brought down the cost of cloth, which stirred consumption activity even deeper into the social order. Fustian in particular found ready buyers in England.[15]

This was a preindustrial world, but it was not precapitalist. While only a few people engaged directly in business beyond their local communities, the existence of markets, merchants, credit, and long-distance trade reveals investments risked in hope of returns. The division of labor, too, appears in the proliferation of guilds in the towns and cities: in the textile business of York alone, dyers, fullers, and weavers occupied distinct organizations by 1415, as did shearmen, wool packers, card makers, and coverlet weavers. Even in the countryside, textile production had already begun to divide at a scale larger than the individual household – government regulations, intent on preserving the reputation of exported products, reveal some of this. The 1196 Assize of Measures, for example, forbade particular finishing operations from the countryside, and limited dyeing to the cities and boroughs. Farm families may have woven cloth from fleece, but finishing it fell to urban dwellers, probably directed by the guild merchants who knew what distant consumers wanted. This division of labor could be viewed as proto-industrialization – after all, the decree dictated that some tasks had to be performed outside the home. But the Assize also specified the width of cloth offered for sale, and the Magna Carta reiterated the requirement in 1215 – trying, in medieval fashion, to ensure standard measures and the quality and reputation of the product.[16]

The 1196 Assize stipulated that specific tasks take place in different locations, and some of these tasks were even mechanized in the medieval period. The heavy work of fulling came out of the home by the end of the thirteenth century when waterwheels were used to raise and drop fulling stocks (giant hammers, larger than an adult person) that pounded cloth

---

[15] John Styles, "Indian Cottons and European Fashion, 1400–1800," in Glenn Adamson, Giorgio Riello, and Sarah Teasley, *Global Design History* (London: Routledge, 2011), 37–46, esp. 39–40; Giorgio Riello, *Cotton: The Fabric That Made the Modern World* (Cambridge and New York: Cambridge University Press, 2013), 73–75.

[16] Heaton, *Yorkshire Woollen and Worsted Industries*, 31–32, 126; Derek Hurst, *Sheep in the Cotswolds: The Medieval Wool Trade* (Stroud, Gloucestershire: History Press, 2005, 2014), end of chapter five; Lee, *Medieval Clothier*, 161.

and felted the wool. Lords could rent the waterfalls to millers and fullers, who collected payment from the tenants bound to grind their flour and full their cloth at the landowners' mills. By waiting for payment, fullers extended credit to clothiers – a financial maneuver that aided the movement and transformation of capital. It was also relatively easy to transform a mill from one purpose to another, from flour grinding or sawmilling to cloth fulling, for example. The replacement of millstones with fulling stocks required some new shafts and gears, but the work could be done by local carpenters or millwrights. Over the years, these mills became the "natural site for other devices introduced to the industry," and the usual location of workshops within which those devices operated. Weavers might operate as clothiers who sold fulled and tentered cloth to merchants who arranged for still further finishing, in line with what the drapers wanted. The increasing divisions of textile processes meant markets between them, too – especially the market for yarn. Yarn spun in one home was sometimes commissioned or purchased by a merchant for weaving in another place.[17]

In other words, long before the Industrial Revolution separated work and production from homes and consumers in recognizable ways, the process had already begun. Mechanization sometimes accompanied these changes, as in the heavy work of fulling wool. Some stages of the work remained crafts and some left the home; parts were mechanized while others used familiar tools and skills. In this way, industrialization would find both existing structures to draw upon and also transformations that were already taking place. Vital to understanding these shifts, however, are the wider worlds within which Europeans operated. As with production processes, the trade systems associated with the expansion of Europe in the early modern period had medieval predecessors. Couched within the medieval world were those legal and economic structures and trade networks that would shape eighteenth-century industrialization, even as they changed in the process.

World Trade

One reason for this growing business was the global conflict known to European historians as the Crusades. Begun in 1096 after the Byzantine emperor came to Rome from Constantinople to plead for help against Ottoman Turks, hoping at the same time to mend a forty-year split in the

---

[17] Gillian Cookson, *The Age of Machinery: Engineering the Industrial Revolution, 1770–1850* (Woodbridge: Boydell Press, 2018), 13; E. M. Carus-Wilson "An Industrial Revolution of the Thirteenth Century," *Economic History Review* 11, no. 1 (Oct. 1941): 39–41; Dickenson, "West Riding Woollen and Worsted," 75–87.

Catholic Church, the Crusades consisted of hundreds of years of religious and territorial conflict between Europe and the Middle East, between Christianity and Islam. Multiple phases of war included the Moorish occupation of Spain and the voyages of European feudal soldiers across the Mediterranean to recapture the Holy Land for Christendom. One result was an increase in commerce between the two regions – partly from trade and partly from plunder. At the same time, equipping the pope's armies strengthened the rule of local lords over their subjects. So the Crusades buttressed provincial authority even as they accelerated intercourse between Europe and Asia, through the Middle East. The Crusader-held territories of Tripoli and Acre fell in the late thirteenth century, and the Europeans retreated, but the struggle flared up sporadically through the 1400s as the Ottoman Empire continued to expand. European imperialism partly sprang from these adventures.[18]

Schoolchildren learn that the Age of Exploration began in the last years of the 1400s when Portuguese and then Spanish, Dutch, French, and English explorers set sail and found new worlds in Africa, Asia, and the Americas. The lands they found were new only to them, however. Wild shorelines and bustling ports were often already linked together by trade in luxury goods, valuable for their bulk and weight, which repaid the effort and costs of shipping or carrying long distances. Valuable items from Asia had for centuries come all the way to Europe on a varied combination of overland, river, and sea voyages now called "The Silk Road." The Silk Road was really many routes and exchange networks, each with its own history. An overland version passed from China through Central Asia, and Persia (now Iran), to the Roman territories of Europe; the sea route linked East and West through the crossroads city of Constantinople. Ships from the Indian Ocean could also sail up the Red Sea to unload merchandise bound for Suez and Alexandria, and on into the Mediterranean Sea. Other ships carried Asian artifacts to the Horn of Africa, for shipment to Ethiopia. The Islamic Middle East, known to European merchants as the Levant, served as a funnel for transshipping Asian textiles from the Silk Road into Europe via the Mediterranean. The merchants of Venice distributed Levantine trade goods across Western Europe.[19]

---

[18] Jonathan Phillips, *Holy Warriors: A Modern History of the Crusades* (New York: Random House, 2010); Jonathan Riley-Smith, *The Oxford Illustrated History of the Crusades* (Oxford and New York: Oxford University Press, 1995).

[19] Debin Ma, "The Great Silk Exchange: How the World was Connected and Developed," in idem., *Textiles in the Pacific, 1500–1900* (Oxon: Ashgate Press, 2005), 1–10; Marika Sardar, "Silk Along the Seas: Ottoman Turkey and Safavid Iran in the Global Textile Trade," in Amelia Peck, ed. *Interwoven Globe: The Worldwide Textile Trade, 1500–1800* (London: Thames and Hudson, 2013), 67–68.

The Silk Road was named for the Asian textile that English sumptuary laws reserved for upper-class consumers, which had reached Europe from at least the time of Christ. Silk was lustrous, desirable, valuable for its bulk, the kind of luxury good that consumers liked on both ends of the Eurasian continent. But the Silk Road carried more than silk. Crockery, especially white dishware, was produced plain on a large scale in China and decorated there in blue patterns to suit specific markets. Blue and white porcelain from China circled the globe into the sixteenth and seventeenth centuries. Spices also traveled through these exchange networks from Asia to Europe. Pepper and nutmeg flavor food and also can be used to preserve it, as salt can preserve raw meat. Exotic spices were also components of drugs, medicines, and perfumes. As a result they were in high demand, among those Europeans who could afford to acquire them from the East where they were grown, through the networks of merchants that used the routes of the Silk Road.[20]

During the Crusades, medieval Europe became more modern as luxurious goods including china, spices, and cloth moved more often across Eurasia. With the artifacts and commodities of the East came ideas and technologies related to their production, distribution, and use. Then, in 1453, Ottoman Turks captured Constantinople from the Christian Byzantine Empire. The fall of the city (from a European perspective) marks for many historians the end of the medieval period, and the start of a new phase of Eurasian contact. From the renamed city of Istanbul, victorious Sultans sought more trade. They expanded the infrastructure that increased overland carriage and communication. Eurasian exchange was sparking a global Renaissance in which objects as "both the outcomes and the conveyors of design" became important expressions of identity, ushering in a consumer age and expanding the reach of the fashion cycle. Thus the voyage of Vasco da Gama from Portugal into the Indian Ocean – the founding moment of the European Age of Exploration – only rearranged commercial and political relations already in existence, undergoing changes of their own.[21]

---

[20] Anne Gerritsen, "Global Design in Jingdezhen: Local Production and Global Connections," in Adamson, Riello, and Teasley, *Global Design History*, 25–33; K. N. Chaudhuri, *The English East India Company: The Study of an Early Joint-Stock Company 1600–1640* (London: Frank Cass & Co., 1965), 5; de Vries, *Industrious Revolution*, 55.

[21] Marta Ajmar-Wollheim and Luca Molà, "The Global Renaissance: Cross-Cultural Objects in the Early Modern Period," in Adamson, Riello and Teasley, *Global Design History*, 12–13; quotation from the editors' introduction, 5; Amelia Peck, "Trade Textiles at the Metropolitan Museum: A History," in her *Interwoven Globe;* and Sardar, "Silk Along the Seas," 66–81.

Merchant Capitalism and Expansion

The Portuguese forced themselves into this history in 1510, seizing the port of Goa on the west coast of India, and fortifying trading posts from Cape Town of South Africa to Japan, with Malacca their entrepôt to exchange Indian cloth for spices.[22] The Dutch surpassed this record in the seventeenth century, even as the English joined in seeking a share of the Indian Ocean spoils. European trading companies (including the Dutch East India Company, founded 1602, and the English East India Company of 1600) granted groups of merchants monopoly rights within their own nation to trade with India. These heavily capitalized trading companies – trading guilds with exclusive rights to some portion of a European nation's overseas business – were usually corporate bodies composed of merchant shareholders, individuals who possessed shares in the Company, which they owned jointly together. These joint-stock companies were financial devices for capitalizing trade, to kit out ships and buy things overseas to sell elsewhere. Members could transfer their shares by sale or bequest, which made the joint-stock company a legal entity that shared certain traits with corporations. Both were collectives: institutions grown from regulations that allowed their investors to share risks and rewards, to multiply their efforts so that any specific setback need not destroy the whole enterprise.[23]

European schemes to mingle in the trade of the Indian Ocean became imperial conquest with the help of an ideology known to its later critics as "mercantilism," or merchant capitalism. Mercantilist theories explained the role of commerce in the political economy of the emerging nation-state and the support each offered the other. Early modern European thinkers assumed there were limits on the wealth of the world. The balance of trade therefore animated their ideas, as each nation sought the biggest portion of the world's riches at the expense of the rest. Mercantilists perceived the goal of commerce to be the acquisition of precious metals by the home country – which it should gain by selling items that others wanted – getting their gold in exchange. Gold was specie, sometimes called treasure. Saleable articles could be made at home or brought from overseas, where they were plentiful and cheap, to vend where they were rare and dear. Mercantilism therefore implied the separation of raw material production from processing and consumption.

---

[22] John Guy, "'One Thing Leads to Another': Indian Textiles and the Early Globalization of Style," 13–14, in Peck, ed. *Interwoven Globe.*

[23] Sanjay Subrahmanyam, *The Political Economy of Commerce: Southern India 1500–1650* (Cambridge and New York: Cambridge University Press, 1990); Harris, *Industrializing English Law,* 24.

It idealized extracting raw materials from colonial locations, adding value by treating or transforming them in the home country, and finding or founding markets to unload the finished goods. Colonies in these networks served as outposts of trade, extraction, and consumption of goods from the mother country. Purchasing cheap goods abroad to re-export to the colonies, or to trade for gold, made good sense to merchant capitalists.[24]

Mercantilist ideology assumed that goods were best manufactured or finished in European workshops, to add value to the colonial raw materials. Mercantilists therefore sought to replace imported consumer goods with substitutes that were made at home. A classic case can be found in the blue-and-white dishes that originated in China. The Dutch East India Company alone increased its imports from 50,000 to 200,000 pieces per year by the middle of the 1600s – a figure that reached 43 million pieces by the end of the eighteenth century. An additional 30 million pieces of china were brought to Europe by merchants of other nations. These imports spurred domestic producers to try to compete; the Dutch city Delft housed thirty large workshops making Chinese-style porcelain by 1670. In England, the number of potteries in the county of Staffordshire alone tripled between the 1680s and the 1750s. Staffordshire firms specialized in making fine earthenware dishes cheaper than Chinese porcelain, if not quite as fine.[25] Such import substitution indicated an important cultural shift: ordinary people had more access to goods and bought more of them as Eurasian trade expanded consumer options.

Consumer Revolutions and Kashmir Shawls

European consumers craved Asian spices, and buying and selling Indian Ocean textiles was one way the European trading companies acquired them to sell back home. In fact, spices mattered so much to the mercantilist business model that England's East India Company (EIC) paid its first dividends in pepper. The EIC – a classic mercantilist trading institution, a kind of merchant guild – brought mirrors, inlaid pistols, and delicate, elaborate clocks as tribute when they received permission to trade on the subcontinent from India's Mughal princes. At first, the EIC bought cloth in India only to barter it for more desirable items, especially spices, and most especially pepper. For its members who preferred cash, the Company auctioned off some cloth in London, including

---

[24] Immanuel Wallerstein, *The Modern World-System II: Mercantilism and the Consolidation of the European World-Economy, 1600–1750* (Berkeley and Los Angeles: University of California Press, 2011).

[25] de Vries, *Industrious Revolution*, 130–32; Gerritsen, "Global Design in Jingdezhen."

not only silk but also the calico fabric made of cotton. By 1614, the EIC was trying to figure out which Indian cloths would sell best in Turkey. By 1650, it had shifted strategies toward importing Indian textiles directly to continental Europe, and cotton cloth was becoming the Company's stock-in-trade.[26] Cotton fabrics represented 73 percent of the value of the Company's imports by the mid-1660s, a number that climbed to 83 percent within twenty years. Mercantilists approved when the cotton calicoes and chintz that the EIC imported from India competed against the Flemish, Dutch, and French linen imports to England, while bringing customs revenue to the Crown.[27]

In fact, the Indian Ocean trade that European merchants wished to join was woven together with fabric. India clothed the world that circulated around the Indian Ocean, and its fabulous textiles had markets on either side of that sea. On the east coast of Africa, Indian cotton cloths were traded for slaves, even before European entry into the business. On the other side of the Indian Ocean lay the Spice Islands. The Dutch, master traders, bought cloth in India and sold it to buy pepper and gold in Sumatra; the pepper could then be traded for Chinese goods and more gold, which in turn became silver in Japan – and all these articles could be traded for spices on Java, which the Dutch seized by force and settled as a colony to serve their commerce.[28] National trading companies competed with one another for the resources they wanted, and they did not often shrink from force. They fought against rebels and interlopers, destroyed farms, and murdered and enslaved native populations to achieve their goods. "Violence was omnipresent" in the 500-year history of European expansion, and the great joint-stock trading companies of European merchant capitalism used kidnapping and rape, plunder and punishment, fire and guns, to configure the trade and the existing business structures in their favor.[29]

Queen Elizabeth had first licensed the English East India Company to trade in Asia in order to produce government revenue by taxing the goods

[26] Chaudhuri, *English East India Company*, 140; Om Prakash, "The English East India Company and India," in H. B. Bowen, Margarette Lincoln, and Nigel Rigby, *The Worlds of the East India Company* (Woodbridge, Suffolk: The Boydell Press), 3–6; Jonathan Eacott, *Selling Empire: India in the Making of Britain and America, 1600–1830* (Williamsburg, VA and Chapel Hill: Omohundro Institute and University of North Carolina Press, 2016), 28, 61.

[27] Melinda Watt, "Whims and Fancies: Europeans Respond to Textiles from the East," 82–103," especially p. 86, in Peck, ed. *Interwoven Globe*.

[28] Prasannan Parthasarathi, *Why Europe Grew Rich and Asia Did Not: Global Economic Divergence, 1600–1850* (Cambridge and New York: Cambridge University Press, 2011), 23; Guy, "One Thing Leads to Another," 17–19.

[29] Dierk Walter, *Colonial Violence: European Empires and the Use of Force* (Oxford and New York: Oxford University Press, 2016), quotation at 5.

it imported. During the English Civil War, the EIC took on new functions. The charter it received from Oliver Cromwell authorized it to operate in more Indian territories including Surat, Bombay, and Bengal. After the Restoration in 1660, King Charles II continued to help the Company's business. He produced new charters that endowed the Company with police, military, and diplomatic powers, as well as the right to mint money for trade on the subcontinent. By the eighteenth century, the government of England and the trade of the Company were even more tightly intertwined than at the start: now Parliament financed some EIC operations, and the Company was authorized to enforce its own laws in its trading settlements. This delegated the sovereignty of the Crown to the Company. As a result, the Company functioned as a government in its territories, a branch of England's government in India. For colonized people, foreign powers ruled their lives.[30]

The cloth that came out of India was produced at the household level, using techniques similar to those employed in Europe. As with English wool, preparatory processes made fluffy cotton fiber lie all in one direction before drawing it out to become the equivalent of a sliver, and then a roving, which was then spun into a yarn suitable for weaving. Women spun cotton in India as English women spun wool into yarn. Male weavers headed households that aided their work at the loom, similar to the organizational structure of woolen production in England. Indian weavers generally owned their looms, raw materials, and finished products, and they also lived in villages that both organized textile production and marketed the individual weaver's wares. A weaver might operate on credit supplied by merchants who specified what cloth he would produce and when and for what price, but the merchants owned neither the cloth nor the tools used to make it. The cloth thus made would wend its way through intermediaries to an export merchant who would finish the cloth for consumers. Typical of textile production that accompanied agricultural subsistence, the household formed the basic unit of production; in India, the village itself also structured distribution networks.[31]

The different regions of the Indian subcontinent each made fabrics of different types. As the markets expanded to include larger numbers of European consumers, India's makers and merchants specialized further. Particular villages and regions became world-famous for particular cloth

---

[30] Sudipta Sen, *Empire of Free Trade: The East India Company and the Making of the Colonial Marketplace* (Philadelphia: University of Pennsylvania Press, 1998), 61–68, 80.

[31] Riello, *Cotton*, 61–66; Ian C. Wendt, "Four Centuries of Decline? Understanding the Changing Structure of the South Indian Textile Industry," in Giorgio Riello and Tirthankar Roy, eds., *How India Clothed the World: The World of South Asian Textiles, 1500–1850* (Leiden and Boston: Brill, 2009).

Figure 1.2 The twisted teardrop design now known as Paisley is reputed
to have originated in Persia, though it was most commonly associated
before the nineteenth century with shawls woven in Kashmir from the
fleece of a Tibetan mountain goat. Courtesy of vectortatu/iStock /Getty
Images.

products. Madras was known for its cottons; the Gujarat region was famous
for embroidered or printed cloth. Bengal was a fabled source of beautiful
silks and also delicate muslins, woven of the finest counts of cotton yarns.[32]
Kashmir produced colorful shawls of the softest wool in intricate patterns,
popular across the Ottoman Empire, in Persia, and in Russia. Kashmiri
shawls were royal gifts and prestige items for courtiers. Made from the fleece
of a mountain goat native to Tibet and Central Asia, reputed to have been
for centuries exported only to Kashmir, the fine wool was woven in extraordinary patterns on simple looms. The ornate designs on the Kashmir
shawls often included a twisted teardrop or pear-shaped motif, now called
Paisley, handed down from Persian design (see Figure 1.2). The twisted

[32] David Washbrook, "The Textile Industry and the Economy of South India, 1500–
1800," in Riello and Roy, *How India Clothed the World*; John Gillow and
Nicholas Barnard, *Indian Textiles* (London: Thames and Hudson, 2008), 26; Riello,
Cotton, 66–67.

teardrop motif indicates the interwoven global sources of textile design and production even before industrialization.[33]

The Calico Craze and Triangular Trade

Despite its popularity in Asia and Africa, the adoption of all-cotton cloth was slow and uneven across seventeenth-century Europe. For one thing, supplies were unpredictable, partly due to the fierce competition among textile buyers. The Dutch East India Company (VOC) bought more cotton cloth than did the English, and understood the market better. Cotton consumption in England developed only haltingly; it took time for English taste and Asian production to adapt to one another. Cotton lay among a range of fabric options, and often wore out faster than more commonplace fabrics. English people generally preferred their cotton woven into a linen warp, producing the harder-wearing fustian. The EIC tried several times to flood the London market with inexpensive cotton shirts and shifts, made up to the Company's order in Madras, but it found few buyers in the 1680s and 1690s. Consumers rejected these ready-made garments because they had other options. Undergarments like these required lots of laundering, but provided little by way of style.[34]

Outer layers of clothing, such as gowns and waistcoats, were more for show, and it was in this category that cloth made in India and shipped to Europe and England first appealed to buyers. Fashion seemed to be accelerating in the last quarter of the seventeenth century, along with modernity. Wealthy purchasers craved new color combinations and floral patterns every season, and imports from India fed the rapid changes they desired. Colorful printed cottons imitated fancy silks, and their lower prices allowed their rapid replacement for more fashionable consumers. In the second half of the 1600s, the East India Company began placing orders in India for cloth printed and woven in European patterns, which elites were wearing as loose long robes – banyans or Indian gowns – or using on the bed, or hanging on the wall. Then King Charles II wore a waistcoat fashioned of Indian cloth and showed how the new textile could be worn in a more conventional outfit. By the end of the century,

---

[33] Caroline Karpinski, "Kashmir to Paisley," *Metropolitan Museum of Art Bulletin*, New Series, Vol. 22, no. 3 (Nov. 1963), 116, 119; Chitralekha Zutshi, "'Designed for Eternity': Kashmiri Shawls, Empire, and Cultures of Production and Consumption in Mid-Victorian Britain," *Journal of British Studies* 48, no. 2 (Apr. 2009), 421–23.

[34] John Styles, *The Dress of the People: Everyday Fashion in Eighteenth-Century England* (New Haven and London: Yale University Press, 2007), 88–114, 128–129; Riello, *Cotton*, 95, 115–16, 130–34.

Indian chintz and calico topped European imports of finished textile cloths.[35]

Increasing imports led English wool merchants to fear that a "Calico Craze" was sweeping England in the last decades of the 1600s – that Indian fabric threatened domestic industry. Indeed, India's importance as a cloth producer can be traced in the words used to describe textiles, words that entered the English language along with its cotton cloth. Calico and chintz, khaki and gingham, pyjama and dungaree and shawl, are all words originating in the subcontinent, along with the clothes they describe. Indian cloth was often very bright, with color that stayed fast through the wash, which contributed to its worldwide popularity. English sumptuary laws had disappeared near the start of the seventeenth century, and traditionalists feared that the new fabrics blurred the visual distinctions between social classes. They argued that common folk had historically worn drably colored wool, and now cotton imports from India made it possible for them to look more like the gentry who could afford patterned silk. In seventeenth-century England, however, the highly printed imports were used mostly for home furnishings, especially large bedcoverings decorated with a Tree of Life. Not until the mid-eighteenth century did cotton become fashionable for clothes for large numbers of Britons.[36]

But the Company had customers for Indian cottons across the Atlantic, in the American colonies, which stretched eventually from the sugar islands of the Caribbean all the way to Hudson's Bay in Canada. Cotton cloth woven in India could fulfill mercantilist English dreams by vending elsewhere – including Africa, where it was traded for people as if they were property. From there, English ships packed with Africans in chains, or Asian or European consumer goods, sailed across the Atlantic to the English settlements along the North American coast. Europeans began slave trafficking at about the same time, and in many of the same patterns, as they began to encroach into the textile business, and the two enterprises meshed together. Both Indian cloth and enslaved African laborers were in demand in

---

[35] Parthasarathi, *Why Europe Grew Rich*, 31; Watt, "Whims and Fancies," 86; Riello, *Cotton*, 126, 130–34; Beverly Lemire and Giorgio Riello, "East & West: Textiles and Fashion in Early Modern Europe," *Journal of Social History* 41, no. 4 (Summer 2008): 895.

[36] Beverly Lemire, *Fashion's Favorite: The Cotton Trade and the Consumer in Britain, 1660–1800* (Oxford and New York: Oxford University Press, 1991), 16; Amelia Peck, "'India Chints' and 'China Taffaty': East India Company Textiles for the North American Market," 105, in Peck, ed., *Interwoven Globe*, 86; Styles, "Indian Cottons and European Fashion," 40–42; Styles, *Dress of the People*, 15, 109–27.

British America.[37] Up to about 1750, the coasts that rimmed the Atlantic Ocean bought between 80 and 94 percent of Britain's cotton exports. Cloth sold in Africa alone comprised 68 percent of all British exports, across the eighteenth century. In exchange, Africans exported not only gold and ivory but also human beings, slaves intended to work in European plantations on the other side of the sea.[38]

Slaves from Africa represent the final thread of European economic expansion in the Age of Exploration. From Portuguese sugar plantations on the Canary Islands to the Spanish silver and gold mines in South America, when natives of conquered regions died rapidly, labor supplies were replenished by European intervention into the slave trade of Africa. The Portuguese had begun trading cotton cloth from India for slaves in West Africa by the 1580s. They were soon joined by Dutch, French, and British traders organized into companies chartered or sponsored by their governments. The slave trade netted England £100,000 a year until the 1720s, then increased tenfold until it reached over £1 million annually at the end of the eighteenth century. The British sold cloth to buy slaves. In the early 1700s, much of that cloth was still made from English wool, but cotton from India began to replace domestic manufactures; Indian cotton eventually reached 40 percent of the cloth the British re-exported to Africa. The Dutch also used cloth to buy slaves in West Africa; it formed 60 percent of their trade goods by the 1640s. For France also, cloth bought slaves: half the items the French brought into the slave trade were cloth pieces from India.[39]

People from Africa labored as slaves on New World plantations, which had been developed originally to grow sugar for European consumers. At first, the word "plantation" meant only a colonial settlement planted and intended to survive and grow. The term plantation distinguished these colonies from trading outposts, which the EIC called factories, for the factors or merchants who operated there. Over time, "plantation" came to refer to a site for the cultivation of agricultural commodities using slave labor. The Portuguese developed plantations on Atlantic islands, off the coast of Africa, and used them to grow sugar in the Caribbean. British settlers also used the word to describe their large-scale monocrop enterprises growing tobacco and rice in North America. Plantations were

---

[37] Philip D. Curtin, *The Atlantic Slave Trade: A Census* (Madison: University of Wisconsin Press, 1972); Paul E. Lovejoy, *Transformations in Slavery: A History of Slavery in Africa*, 3rd ed. (New York: Cambridge University Press, 2012).

[38] Joseph E. Inikori, *Africans and the Industrial Revolution in England: A Study in International Trade and Economic Development* (Cambridge and New York: Cambridge University Press, 2002), table 9.9, p. 448; Riello, *Cotton*, 138–40; Eacott, *Selling Empire*, 65–66, 74–79.

[39] Riello, *Cotton*, 137–47; Inikori, *Africans and the Industrial Revolution*.

households where people lived, as the original usage of the word denotes – the planted settlement, as opposed to just a space for trade. However, the large scale of production, the division of labor, and the regularization of tasks and techniques made plantation agriculture a precursor to industrial manufacturing.[40] The intensified cultivation of early modern mercantilist capitalism also provided capital to some early English industrialists, including Samuel Greg, whom we will meet in the next chapter. Finally, when industrialization cohered in the late eighteenth century, the slave-and-plantation system of production would also be adapted to the cultivation of cotton, the characteristic raw material of the Industrial Revolution.

The Indian and Atlantic Ocean trade networks each generated the capital and connections that would later serve the industrialization process. A broad brush could paint one as making sugar while the other made spice. These commodities, and the production and trade networks that both engendered and sprang from them, provide the context for the Industrial Revolution. Spice characterized the European demand for the commerce of the Indian Ocean, though Europeans rapidly extended this impulse into the textile trade around which the business was organized. Sugar plantations supplied capital, and organizational models of routinized tasks and interchangeable laborers. The settlers of the new world developed plantations that enlarged the agricultural household and grew crops for sale and processing in Europe. They bought labor and passed laws creating heritable conditions of slavery in order to serve these mercantilist dreams. And cotton cloth from India anchored both Indian and Atlantic Ocean economies.

Echoes

India's textile industry had its own systems and networks, and these played a role in shaping the development of both European tastes and European business structures. To begin with, the EIC had little control over the quality of the fabric available for purchase, and layers of Indian merchants limited the Company's access to weavers. Nonetheless, increasing European demand contributed to a more complicated division of textile labor that was emerging in India by the 1670s. Spinning and weaving were separating into different locations, which reflected the imbalance in which one weaver required the work of at least four spinners

---

[40] Indrajit Ray, *Bengal Industries and the British Industrial Revolution (1757–1857)* (London and New York: Routledge, 2011), 55; Sidney W. Mintz, *Sweetness and Power: The Place of Sugar in Modern History* (New York: Penguin Books, 1986; orig. pub. 1985), 46–61.

to supply the loom. Although observers later claimed that a similar bottle-neck incentivized the mechanization of spinning in eighteenth century England, this ratio did not have the same results in India a century earlier. Comparing the two cases demonstrates that economic incentives are inadequate to account for technological change, even when they are a necessary element to explain its success. In India, marketing the cloth through village merchants meant that weavers maintained control of their own time and products. The merchants with whom they contracted undertook the finishing that made the product so competitive in world markets. Quality and finishing, printing and painting, and bright dyes that would not soon fade in the wash mattered more than low cost in making Indian textiles attractive around the world.[41]

In fact, Indian cloth entered European textile markets that were them-selves dynamic in the seventeenth century, despite and during the terrors of the Calico Craze. English makers in many sectors had adopted new machinery: in Derbyshire and southern Nottinghamshire, for example, the Midlands industry that knit hose out of fine yarn began mechanizing, using devices known as stocking frames or knitting frames, which multi-plied common work and standardized its results. Spinning wheels also grew larger in the sixteenth century, sometimes reaching five feet tall. A Great Wheel required the spinner to walk to and from the apparatus when drawing out the fiber and twisting it into yarn, which still resulted in lumpy, bumpy yarn – but more of it than spun with smaller wheels. Saxony Wheels were also coming into use, set with a flyer – the U-shaped device called a "Winding Arm" in Figure 1.3, whose pegs twisted the yarn as it was wound onto the bobbin. This flyer made the process more continuous and could double a wheel's productivity.[42]

Sometimes laws suppressed new machinery. Tudor legislation in the mid-sixteenth century, for example, prohibited the use of gig-mills in the final stages of finishing wool. These drums rotated their large iron frames against the cloth, frames filled with teasels (stiff thistles) to raise the nap of the woven, fulled, and tentered cloth in order to use shears to crop it level (see Figure 4.1). These could be powered by falling water, as fulling stocks were. While the prohibition against gig-mills sometimes faltered as the centuries passed, they were not yet widely adopted. Some early adopters found their mills under violent attack, and sometimes royal proclamations outlawed the machine.[43] For hundreds of years, finishing

---

[41] Riello, *Cotton*, 66–80, 107.
[42] R. S. Fitton and A. P. Wadsworth, *The Strutts and the Arkwrights, 1758–1830: A Study of the Early Factory System* (Manchester: Manchester University Press, 1958), 24–25; Dickenson, "West Riding Woollen and Worsted," 61; Lee, *Medieval Clothier*, 47–48.
[43] Lee, *Medieval Clothier*, 57–59.

Two-handed Spinning-wheel.

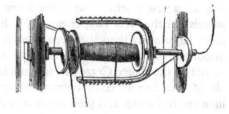

Spindle, Bobbin, and Winding-arm
on a larger scale.

Figure 1.3 Spinning wheel, outfitted with a flyer. Medieval textile technology included the adoption of the Saxony Wheel, whose flyer twisted the yarn as it spun onto the bobbin, at right. This made more and smoother yarn, and was incorporated into the later spinning devices associated with the Industrial Revolution. Courtesy of duncan1890/ DigitalVision Vectors/Getty Images.

wool remained divided in ways that meant some processes, like cropping, were done by hand, while others, like fulling, lodged outside the home and employed heavy machinery. Industrialization was a process that took place in piecemeal ways even before it could be identified as mechanization, the separation of production from consumption, and improvements in the flow of physical materials. These three parts of the definition did not always take place at the same time or in a particular order.

Great Wheels and Saxony Wheels for spinning fleece into yarn, stocking frames for knitting hose, and the forbidden gig-mills for finishing wool pieces, were all new machines. As we shall see, these innovations were even associated with the development of new fabric types, and shifts in the organization and locations of production. But they did not yet spark an Industrial Revolution. Machines did not themselves result in mass production. Fresh fibers in increasing volumes, new markets, and power and labor supplies and structures were not yet available to make new machinery have revolutionary effects. In order to see how these elements became part of the industrial system, we will incorporate the world outside of machines to explain mass production and industrialization. Long-term trends in consumer preferences and the product innovations that met that demand is one more set of important elements to sketch; another is the legislation that would shape market structure as the eighteenth century turned.

New Draperies

The important innovation in the textile business of early modern England was not a new machine but a new fabric. Not the new flyers and spinning wheels but the New Draperies – a product of the sheep's fleece prepared more laboriously than the steps used to make wool – represent the next important part of the Industrial Revolution story. By the time merchants identified the Calico Craze threat in the late seventeenth century, they were already meeting the rising demand for smoother, softer, more colorful and lightweight fabrics, and their products were known generically as worsteds. The thin smooth yarns of worsted used only the longest fibers of the fleece. Several stages of combing and straightening these long fibers separated out the short and curly ones, the noils that combers could then sell to wool producers before the longer, combed fibers were prepared and spun into worsted yarns – softer and less itchy than wool. Both wool and worsted yarns could be combined with cotton, silk, and linen (made from flax plants). Mixtures designed for different purposes eventually included kerseymeres, serges and shalloons, beaverettes and bays, calimancoes, camblets, and tammies. Some were broad and some were narrow, dyed in the wool or undyed. All were product innovations that helped accustom both consumers and producers to new goods with different characteristics.[44]

By the end of the seventeenth century, the center of cloth production of these New Draperies had shifted north in England, which was united with Scotland into the kingdom of Great Britain in 1707. Yorkshire County was doing well with worsted production, benefitting from the New Draperies. As new methods spread into new regions, they bore marks of new business organization too. Yorkshire's worsted makers had more

---

[44] Heaton, *Yorkshire Woollen and Worsted Industries*, 3–7, 217; Dickenson, "West Riding Woollen and Worsted," 60, 116; Hudson, *Genesis of Industrial Capital*, 26–27.

capital than wool clothiers, and operated on a bigger scale from the start. They had to invest more money in raw materials through the more complex processes of preparing and spinning, and they divided the labor into units beyond the size of the usual household earlier than woolens clothiers did. There were medieval predecessors; just as with the Industrial Revolution, the changes in business configuration and methods that accompanied the production of worsteds extended beyond the time period most associated with their technological transitions.[45] But the rise of the New Draperies in the sixteenth century and the movement of commercial cloth production to Northern Britain also trace the decay of a more medieval way of doing business, from guilds to fulling wool pieces at the lords' mills.

In the West Riding of Yorkshire, for example, in the countryside around Halifax, farms divided among children over several generations resulted in smaller and smaller holdings that could not support whole families, who therefore relied on textile production to make ends meet. As a result, the family division of domestic labor became specialization of function in the sixteenth century: spinning and weaving worsteds were often tasks performed in separate households. Some families spun yarn and others wove those yarns, and merchants carried the goods from one to the other. These regional historical contingencies suited merchants prepared to invest more money in specialized tasks, and divided-up production, to make the new cloth types. In contrast, woolen fabrics from the more self-sufficient farms in the eastern part of the county, around Leeds, were still made within older structures. Weavers still used family labor and also still farmed or at least participated in the harvest. Some of them, especially the more substantial clothiers, even trained apprentices in the craft.[46]

The separation of weaving and spinning into different households, managed by merchants who carried goods from one house to the next, indicate that the domestic system of production was changing. Scholars call this system that made the New Draperies proto-industrialization, or the "putting-out system." Merchant-manufacturers put raw materials out into people's homes for processing: they managed cloth production, and paid other people to do the work, separating work into tasks that could be done in different locations, which they coordinated to make cloth. As early as 1555, whole areas concentrated on making a specific component of cloth. For example, combing the longer fibers out of the fleece to be made into worsted yarns was a specialist profession in some areas.

---

[45] Heaton, *Yorkshire Woollen and Worsted Industries,* 3–7, 297–98; Dickenson, "West Riding Woollen and Worsted," 31–34; Lee, *Medieval Clothier.*
[46] Hudson, *Genesis of Industrial Capital,* 26–27, 62–64.

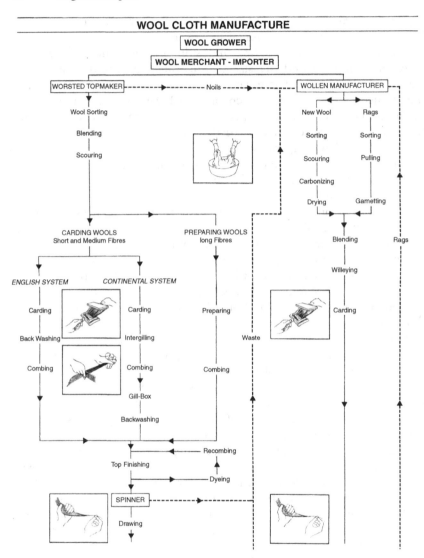

Figure 1.4 Preparing fiber for spinning woolens and worsteds. Courtesy of Bradford Industrial Museum, Museums and Galleries, Bradford Metropolitan District Council.

Combing for several customers could maintain a household's independence, while some worsted combers were also weavers who wove woolen cloth out of the noils discarded from the longer fibers during the combing

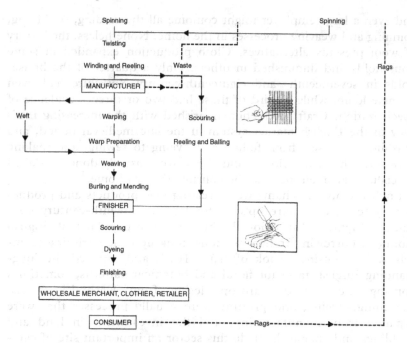

Figure 1.4 (cont.)

of worsted. Even making the combs used by worsted combers had become individual employment out of more general blacksmithing shops. Spinning, on the other hand, was more widely dispersed across the region's households, and a weaver might himself put out spinning to many households. Some even paid out-weavers and supplied their looms, warps, and wefts.[47]

Meanwhile, woolen production also expanded into new contours in the quickening commerce of early modern England. By the mid-1600s, some of the wealthier woolens makers were establishing workshops. They might put a second story on their homes to house their looms. Or they could build extensions for weaving work, often equipped with wide windows to illuminate their operations. Some used the changing architecture to house more than one loom, and sometimes broader looms. Masters who employed apprentice weavers expanded their households to accommodate them. However, it was still common to train them in the whole process of cloth manufacture,

---

[47] Smail, *Origins of Middle-Class Culture*, 22, 53; Hudson, *Genesis of Industrial Capital*, 59–65; Dickenson, "West Riding Woollen and Worsted," 116–23.

and even a large employer might combine all the carding, scribbling, spinning and weaving processes in the home. Nonetheless, the history of wool presents alternatives. Cloth production expanded in some households and diminished in others. Only a quarter of the households in seventeenth- and eighteenth-century Yorkshire had even a single loom, while a third of those had two or three – evidence of specialization. Craft production flourished with the increasing intricacy of the English market system in the late medieval period, and finishing processes, from fulling and dyeing to sewing, and making accessories like hats, gloves, and shoes, were usually done by skilled specialist craftsmen, outside the consumer's own home.[48]

English cloth merchants were creating new products and production systems, and more options for late seventeenth-century consumers. Light, bright, colorful fabrics were so popular that English clothiers, borrowing Indian methods, took up calico printing themselves. The business took off after 1670, and satisfied the long-standing English taste for floral and botanical patterns, sometimes copying the twisted teardrop design from Kashmiri shawls. Bleaching, dyeing, and printing were usually processes that were separate from manufacturing, and the investments in land and buildings and chemicals made this sector an important site of capital accumulation in British textiles. Water was crucial to these undertakings, and printing found a convenient home along streams in the north, especially in Lancashire County. Some domestic clothiers simply adopted the name of the competition: "Manchester cottons" were actually coarse wool pieces from Lancashire, a county whose denizens also made a lot of fustian by the eighteenth century.[49] They got linen warps from Ireland and Germany, while raw cotton fiber arrived mostly from the West Indies islands in the Caribbean, which exported nearly a million pounds of the fiber in 1697–1698. Mercantilist theorists lauded Manchester for the region's ingenious cotton blends which added value to the imported raw material.[50]

[48] Dickenson, "West Riding Woollen and Worsted," 61–87; Carole Shammas, *The Pre-Industrial Consumer in England and America* (Oxford: Clarendon Press, 1990), 27–33; Masschaele, *Peasants, Merchants, and Markets.*

[49] Geoffrey Turnbull, *A History of the Calico Printing Industry of Great Britain*, ed. John G. Turnbull (Altrincham: John Sherratt and Son, 1951), 11–12; Edward Baines, Jr., *History of the Cotton Manufacture in Great Britain* (London: H. Fisher, R. Fisher, and P. Jackson, 1835), 91–94; Styles, "Indian Cottons and European Fashion," 43.

[50] Alfred P. Wadsworth and Julia De Lacy Mann, *The Cotton Trade and Industrial Lancashire, 1600–1780* (Manchester: Manchester University Press, 1931), 13–15;

Calico Acts

New Draperies and fresh business structures did not relieve the anxieties of England's clothiers, however, and they hounded Parliament throughout the 1690s for some protection against the fearsome Calico Craze. In the new century, Parliament gave them what they wished for: a series of prohibitions against wearing imported printed fabrics, a restriction within which to nurture and indulge the maturing cloth industry. Beginning in 1701, and renewed twenty years later, the regulation against colorful Indian imports spurred innovations in the interstices of the law. The 1701 Calico Act still permitted the importation of undyed cotton fabric, and printers pounced. They paid import duties on the plain fine imported fabric but printed it in England and found customers both at home and across the Atlantic. Demand inspired further efficiencies in the dye and printing sector, and the dropping prices of British-printed imported fabric infuriated traditional clothiers. Even the East India Company complained about being undercut in the export markets by goods printed more cheaply at home. The 1721 Calico Act forbade wearing *any* printed all-cotton calicoes, so the printers learned to make do with mixtures, though fustians rarely took color as well. The prohibitions were specific enough that it was possible to evade them with slightly different wares, and "Callico Madams" were castigated in popular culture as immoral women, usually domestic servants with the power to ruin the national economy with their clothing choices. Violations sometimes met violence, as when rioting silk weavers in London assaulted women suspected of wearing calico garments. They splattered them with acid, and tore the offending clothes off their bodies.[51]

Due to the Calico Acts, then, the British textile industry of the eighteenth century developed behind walls of market protection, and at the same time met considerable competition on distant shores and in colonial markets. Production systems likewise combined both novel and well-known organizational forms, household production

Henry Smithers, *Liverpool, Its Commerce, Statistics, and Institutions with a History of the Cotton Trade* (Liverpool: Thomas Kaye, 1825), 119–20.

[51] Chloe Wigston Smith, "'Callico Madams': Servants, Consumption, and the Calico Crisis," *Eighteenth-Century Life* 31, no. 2 (Spring 2007), 30–34; Turnbull, *History of the Calico Printing Industry*, 11–25; Beverly Lemire, ed., *The British Cotton Trade, 1660–1815*, 4 vols. (London: Pickering & Chatto, 2010), 3:5–11; Patrick O'Brien, Trevor Griffiths, and Philip Hunt, "Political Components of the Industrial Revolution: Parliament and the English Cotton Textile Industry, 1660–1774," *Economic History Review* 44, no. 3 (Aug. 1991): 398–410.

and putting-out merchants, side-by-side in the woolen and the worsted industries, sometimes even within the same firm. For example, William Thomas was a weaver who had five looms in 1714, and lots of raw wool at home and more put out to spinners across Lancashire who spun the yarn for his looms. His three sons wove as he did, and other men off-site also wove fabric in their homes. He employed seven tenters and finished his own cloth. He was not the only example of this system. Men who owned looms in other weavers' homes illustrate the very definition of industrial capitalism, as they profited by investing in tools (the means of production) that other people operated for pay. The putting-out system, however, had developed from the older economic structures of domestic production – as industrialization later would draw on existing forms even as it transformed them. Some apprentices still expected to learn farming as well as weaving, which shows how the old combination of farming plus crafts persisted into the proto-industrial putting-out era. At the same time, prosperous clothiers reflected their wealth through landholdings and agricultural investments. Industry was not yet thoroughly distinct from agriculture.[52]

The putting-out system of production, a major transition in the history of capitalism, took place mostly in areas outside of guild control. Putting-out sprang from domestic production and its roots were dug in country soil. Nonetheless the putting-out system demonstrates that the social relations of rural England were also changing. In the second half of the eighteenth century, a wave of enclosure combined many small farms into large ones and in many places prevented the use of common fields for grazing sheep. Tenant farmers were sometimes left without their ancestral employment. They wandered the roads, seeking work, at the same time that population was growing, agricultural productivity was increasing, and markets and towns were thriving.[53]

Eighteenth-Century Distribution and Specialization

Distribution networks were also shifting with the new patterns of international trade. As the New Draperies flowed from the evolving

---

[52] Smail, *Merchants, Markets, and Manufacture*, 113; Dickenson, "West Riding Woollen and Worsted," 74–98.

[53] Robert C. Allen, "Tracking the Agricultural Revolution in England," *Economic History Review* 52, no. 2 (May 1999): 209–35; Craig Muldrew, *Food, Energy, and the Creation of Industriousness: Work and Material Culture in Agrarian England, 1550–1780* (Cambridge and New York: Cambridge University Press, 2011), 275, 320.

production systems in the North, cloth markets in Lancashire and Yorkshire expanded and began to specialize. In Leeds, for example, a lot of cloth business was done on the bridge that marked the town's boundary. In the eighteenth century, the textile market had a reputation on the Continent, and it had also expanded up the Briggate to the more general provisions market. Two mornings a week, trestles were brought out to line the bridge and the streets, and a clothier might bring a single piece of cloth to display on the boards for sale. A bell opened the market, and merchants walked up and down, examining the wares. Deals were struck in whispers, as both transactors preferred to keep their price negotiations quiet. Within an hour-and-a-half the cloth was sold, or taken away to try for a sale on another day. Another bell marked the close of market and the whole operation shut down – clothiers retreated and trestles were removed.[54]

At Leeds market in the eighteenth century, a buyer might be making purchases on behalf of a London firm, while European merchants sometimes sent their own representatives, and some had particular colors or patterns they wished to match. They could walk up and down the trestles and hold their "foreign letters of order" up against the available cloth pieces, to compare supply with demand. From the bridge at Leeds a piece of wool from the Yorkshire countryside could go anywhere. Strings of packhorses, nimble animals laden with goods, carried bales of cloth pieces where carts could not go. Tiny bridges over the region's small brooks and becks were sometimes the only roads needed to bring the fabric out to consumers who would use it. Packhorse transport survived into the eighteenth century even when towns built canals to link with their hinterlands. The Aire-Calder Navigation, for example, improved Yorkshire's river junctions in the first years of the 1700s, and eventually connected Leeds to the sea, to Hull, and to York itself by 1732. Canal improvements linked northern cloth production to the rest of the world. Leeds in the eighteenth century was second only to London in British sales of woolen fabrics. As the century matured, some Yorkshire merchants got the goods over the Pennine mountains into Lancashire, to the port at Liverpool,

---

[54] Daniel Defoe, *A Tour Thro' the Whole Island of Great Britain, Divided into Circuits or Journeys*, 4 vols., 4th ed. (London: S. Birt etc., 1748), 3:97–121; Herbert Heaton, "Leeds White Cloth Hall," *Publications of the Thoresby Society* 22 (1915), 134–35.

whence they shipped goods directly to North America, a crucial source of demand.[55]

The Leeds cloth market outgrew its bridge, and the street was too small to accommodate both the trestles of cloth pieces and the usual traffic that wished to pass. Meanwhile both Halifax and Wakefield were opening markets that might draw buyers away, so the Leeds mayor joined the merchants and got a Cloth Hall open to sell goods in 1711. Soon it was too small. New halls were regularly built and improved, until finally the last Cloth Hall opened in 1775, a portion of which still stood in Leeds in 2018, behind the Corn Exchange. Some Leeds merchants apprenticed their children to overseas firms just to acquaint them with the specifics of that demand and mode of doing business, while other Yorkshire merchants lived or sent representatives abroad.[56] These increasingly specialized distribution networks helped regularize the flow of goods between supply and demand. Merchants and makers sometimes overlapped. Some factors were so precise in their needs that they slipped into involvement in manufacturing by ordering custom-made goods or investing in production – especially in the highly capitalized proto-industrial worsted line. A large merchant could keep men at their looms and buy all they made, just to have ready when orders arrived. A clothier might bypass the Halls, or bespeak cloth made to his order in specific colors and patterns. Yorkshire merchants in the mid-eighteenth century sometimes bought generic goods and then dyed, printed, and finished them to meet the buyer's demands.[57]

This wide variety of systems for producing and distributing many types of cloth was a feature of the British textile industry in the eighteenth century. A mature, dynamic, cloth-producing sector responded to the competition presented by imported cotton cloth from India by making new mixtures and reorganizing production, distribution, and marketing methods into new locations and institutions. These shifts and innovations occurred under the wing of government protection in the eighteenth-century Calico Acts. At the

[55] Dickenson, "West Riding Woollen and Worsted," 168–89, 259; R. G. Wilson, *Gentlemen Merchants: The Merchant Community in Leeds, 1700–1830* (New York: Augustus M. Kelley and Manchester University Press, 1971), 81–82; Smail, *Merchants, Markets, and Manufacture*, 91–93, 115.

[56] Heaton, "Leeds White Cloth Hall," 135–37, 139. Description, Papers of the Leeds White Cloth Hall, Brotherton Library Special Collections, University of Leeds, UK, http://library.leeds.ac.uk/special-collections-explore/6443, accessed 11 September 2015; Dickenson, "West Riding Woollen and Worsted," 126, 168–78.

[57] Smail, *Merchants, Markets, and Manufacture*, 63–68; Dickenson, "West Riding Woollen and Worsted," 170–82.

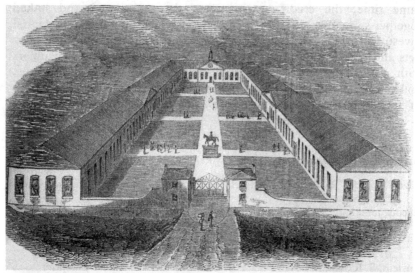

Figure 1.5 Cloth Halls, like this one built in eighteenth-century Leeds, were markets for the exchange of cloth that was usually woven in the countryside around town. Enclosed cloth halls limited access to those who had privileges to trade in town. They superseded the old markets for cloth that had taken place twice a week, on the bridge and up the road, in the 1740s. The big windows of the Cloth Hall provided light to judge the goods. Courtesy of Universal History Archive/Universal Images Group/Getty Images.

same time, the credit arrangements that supported and linked together so many sections of this trade also indicate increasing capital investment in raw materials and equipment. Protected domestic markets and competition abroad; the fashion cycle, and more consumers with more money to spend; innovation in products, raw materials, and the organization and distribution of production; shifting patterns of trade on the world stage: these elements gave British textiles a strong base from which to grow.

Suggested Readings

Clark, Gregory. *A Farewell to Alms: A Brief Economic History of the World.* Princeton and Oxford: Princeton University Press, 2007.

Eacott, Jonathan. *Selling Empire: India in the Making of Britain and America, 1600–1830.* Williamsburg and Chapel Hill: Omohundro Institute of Early American History and Culture and University of North Carolina Press, 2016.

Epstein, S. R., and Maarten Prak, eds. *Guilds, Innovation and the European Economy, 1400–1800.* Cambridge and New York: Cambridge University Press, 2008.

Long, Pamela O. *Openness, Secrecy, and Authorship: Technical Arts and the Culture of Knowledge from Antiquity to the Renaissance.* Baltimore: Johns Hopkins University Press, 2001.

Mintz, Sidney W. *Sweetness and Power: The Place of Sugar in Modern History.* New York: Penguin Books, 1986.

Ogilvie, Sheilagh. *Institutions and European Trade: Merchant Guilds, 1000–1800.* Cambridge and New York: Cambridge University Press, 2011.

Parthasarathi, Prasannan. *Why Europe Grew Rich and Asia Did Not: Global Economic Divergence, 1600–1850.* Cambridge and New York: Cambridge University Press, 2011.

Peck, Amelia, ed. *Interwoven Globe: The Worldwide Textile Trade, 1500–1800.* London: Thames and Hudson, 2013.

Riello, Giorgio, and Prasannan Parthasarathi, eds. *The Spinning World: A Global History of Cotton Textiles, 1200–1850.* Oxford and New York: Pasold Research Fund and Oxford University Press, 2009.

Subrahmanyam, Sanjay. *The Political Economy of Commerce: Southern India 1500–1650.* Cambridge and New York: Cambridge University Press, 1990.

2   Myths and Machines

Peopled with heroes, myths serve a purpose. The word implies falsehood and fable, but myths express deeper truths. They may explain origins, or convey cultural values, much as the Garden of Eden roots its believers in a dream of paradise lost. Even true stories can achieve mythic proportions when they express important lessons because myths reveal how a culture perceives itself and its world. Invention myths are no exception. They frame our understanding of the Industrial Revolution. In the mythic version, John Kay's flying shuttle, patented in 1733, accelerated weaving and thereby strained the domestic model of production and the house-hold supply of spun yarn for the loom. The flying shuttle therefore inspired a series of new machines for spinning: James Hargreaves' spin-ning jenny and Richard Arkwright's water-frame, both patented in 1769, followed ten years later by Samuel Crompton's unpatented spinning mule, which combined the approaches of the 1769 machines. In this story, this sequence of ingenious spinning machines caused industrializa-tion. This list of innovations possesses mythological power because it communicates important cultural lessons: that necessity is the mother of invention; that geniuses solve problems with clever devices in a flash of discovery, and their machines change the world. Ascribing technological transformations to a series of individual men and machines provides an easy narrative of progress toward mass production, as if mechanization itself caused the separation of consumption from production and the flow of raw materials.

Yet myths are never the whole story. A catalogue of inventions does not quite explain why they worked then and there when the economic incen-tives had long been operative, or when versions of most of the machines had existed for some time – some, for centuries. How were these devices fitted into existing systems, and what networks of people and material objects kept those arrangements operating, and who changed which parts of the whole when component parts changed? Instead of tracing a sequence of improvements, historians of technology usually try instead to explain why those machines were made to work then, in that time and

57

place, and what rearrangements got them running. As a result, this chapter takes a contrapuntal form. It tells the usual stories about the mechanization of spinning but also picks them apart to see how entangled with political machinations and social and commercial contingencies these technical changes were. Ultimately, this chapter seeks to replace one meaning of the word "genius" with another. Instead of the exceptionally talented individual, genius here reveals an alternate meaning: the distinctive, defining characteristic of an age, the animating spirit of a person or a thing, its essential atmosphere.[1] Probing traditional stories more deeply permits an observer to see behind the individual genius to the contextual genius his work expressed. Later, we shall find myths in formation, and investigate their purposes, and replace them with these more complexly woven stories of how technologies change.

To begin, then, with the usual story: John Kay's 1733 flying shuttle employed a piece of string to automate one of the steps involved in weaving. Previously, the weaver had used his hands to pass the shuttle through the shed to make one pick of the weft, one thread of a woven fabric. In John Kay's patented engine for "opening and dressing wool," the weaver instead tugged a piece of string (at P in "Mr. Kay's Lathe" in Figure 2.1) that sent the shuttle flying along a track. The shuttle carried the weft yarn through the shed – the space created when alternate warp yarns (B in "The Loom" in Figure 2.1) were lifted by one of the heddles (C in the same figure). Stepping on a treadle at H lifted the other heddle, lifting other warp yarns, making a new shed. Then, tugging the string sent the shuttle – with its weft yarn – over the warp yarns it had just passed under. Tug a peg, rather than pass the weft by hand over the whole width of the warp: this sped up weaving and also made it broader – no longer limited to the span of a man's arms. Before the shuttle was made to fly, cloth wider than an arms' breadth required two weavers, or a weaver and his son, or two apprentice boys, stationed at either end of the loom to toss the shuttle back and forth. Before, even sails were woven only two feet wide, then sewn together. John Kay's flying shuttle carried the weft faster and further. It automated the results of the weaver's pull of the string and multiplied his actions.[2]

---

[1] Oxford English Dictionary Online, s.v. "genius, n. and adj." esp. definitions 6–9, accessed 4 August 2018.

[2] John Kay, "New Engine or Machine for Opening and Dressing Wool," English patent 542, granted 1733; John S. Lee, *The Medieval Clothier* (Woodbridge: Boydell Press, 2018), 50–51; Richard Sims, *Sailcloth, Webbing, and Shirts: The Crewkerne Textile Industry* (Bridport: Somerset Industrial Archaeological Society and the Gray Fund of the Somerset Archaeology and Natural History Society, 2015), 15.

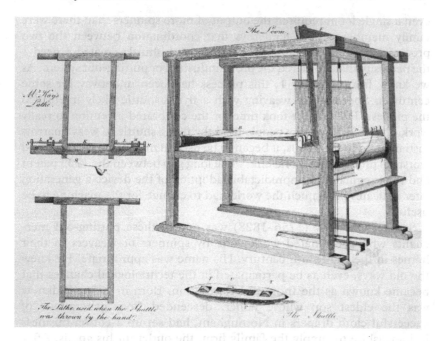

The Loom.

Mr Kay's Lathe.

The Lathe used when the Shuttle was thrown by the hand.

The Shuttle.

Figure 2.1 John Kay's flying shuttle was part of a 1733 patent for "Opening and Dressing Wool," and included the track D, down which the shuttle flew after being tugged into action by the weaver at G. Later writers said that this innovation increased the speed of weaving and so needed more spun yarn than ordinary families could supply. This bottleneck has been given as the reason that innovations in spinning began to appear about thirty-five years later. Photographed by the author from Richard Guest, Compendious History of the Cotton Manufacture (1823).

According to the story, the flying shuttle sped up weaving and thus created what economists call a bottleneck – unmet demand for spun yarns. By 1770, as a result of the flying shuttle, each English weaver needed between five and eight people spinning to keep him supplied with yarn. Historians have found a "distinct connection" in Yorkshire between usual size of families in the area and the labor needed to produce the local cloth, and faster looms would have broken that link.[3] Because

[3] William Radcliffe, Origin of the New System of Manufacture, Commonly Called "Power-Loom Weaving," (Stockport: James Lomax, Advertiser-Office, 1828), 59–60; Pat Hudson, The Genesis of Industrial Capital: A Study of the West Riding Wool Textile Industry, c. 1750–1850 (Cambridge: Cambridge University Press, 1986), 63.

even a single loom required the output of more spinners than there were family members, the legend says that coordination between the two processes became crucial, and thus the flying shuttle created a need – the necessity that mothered the proto-industrial or putting-out system. As we know from Chapter 1, this process had been underway for some centuries. Speeding up weaving with a flying shuttle likely intensified the process. However, it took time for the celebrated invention to really work. In the 1750s, some people used the flying shuttle to weave narrow fustians, and in the 1760s, it began to be used in making the woolens and worsteds produced in Yorkshire.[4] The long gap between the 1733 patent and the irregular and unpredictable adoption of the device a generation later indicates how much the world had to change for the machine to be useful.

Samuel Oldknow (1756–1828) was one of these putting-out merchants who coordinated work done by spinners or weavers in their homes in the eighteenth century. His name was appropriate. He knew the old ways, even as he participated in the technological changes that became known as the Industrial Revolution. Born in 1756, Oldknow was the eldest son whose father, descended from generations of successful cloth drapers in Nottingham, had set up a cotton business in Lancashire to supply the family firm, the outlet for his goods. After a grammar-school education, Samuel would follow in his father's footsteps and pursue cotton manufacturing in Lancashire in the late 1760s. The timing was fortuitous. At first, Oldknow produced cotton cloth using putting-out arrangements, but when he invested his capital in mills and machines, he became one of the early investors in the factory mode of production. He partnered for a time with Richard Arkwright, the most famous industrialist of the age, and in the 1780s Oldknow was renowned for making very fine cotton cloth that could compete with Indian muslins. Samuel Oldknow's career traces the transformations of industrialization: the machines, the business structures, and the social context.[5]

Before Samuel Oldknow industrialized his operations, his putting-out business consisted of carrying the fiber through multiple phases of processing. Workers in their homes picked apart the raw cotton, discarding the seeds and stems and trash; others engaged in carding

---

[4] Geoffrey Timmins, "Technological Change," in Mary B. Rose, ed. *The Lancashire Cotton Industry: A History Since 1700* (Preston: Lancashire County Books, 1996), 39; Michael Dickenson, "The West Riding Woollen and Worsted Industries, 1689–1770: An Analysis of Probate Inventories and Insurance Policies," (Ph.D. Diss.: University of Nottingham, 1974), 61.

[5] Handlist, Samuel Oldknow Papers, John Rylands Library, University of Manchester, UK.

and roving through spinning, and eventually the weaving that turned spun yarn into cloth. Although he managed the entire production process, from fiber to finished cloth, he was a merchant, since his money lay entirely in the raw materials and not in any equipment. He added value simply by moving the fiber through production. The division of labor organized his accounts – he kept different books for picking and winding and spinning and warping (stringing the warp yarns, to prepare them for the loom) and, of course, weaving. Within a book, he would organize the workers according to their locations – many of his pickers lived in Longhouse Lane, for example. Picture him stopping in at one door and then the next, collecting the picked-over, cleaned fiber to give to someone else to card and spin, paying the workers, and providing raw materials for the next period of work. From a family of drapers and a father in cotton, Oldknow became a classic version of the putting-out merchant. He used existing forms of household production, even as he moved goods from house to house, paying for the work that increased its value.[6]

In retrospect, this proto-industrial way of business presented merchants with difficulties. Oldknow's investment swelled as he moved the commodity along the chain. With value increasing at every stage, his initial purchases compounded in worth and tied up ever more of his money. Yarn itself formed half the cost of finished fabric, so he had a lot invested even before weaving and finishing began.[7] Quality control was essential for making fine muslins, but tricky to manage with a dispersed workforce, even if so many pickers lived in one lane. Thievery was always a lingering question. Sometimes pickers made more waste than usual, generating more "trett" and less "neat" fiber than he expected. Should he blame the picker or the raw material he had supplied?[8] Even an honest worker could stop working once she had earned enough for now – or when the raw materials ran out. It was typical of putting-out that workers had little incentive to work beyond their immediate needs, and many had alternate occupations. Farming came first, for many families, and yarn or cloth for the trader might have to wait. Increasing production set its own problems – how

[6] "Pickers book, June 1792–Jan. 1793," Samuel Oldknow Papers.
[7] George Unwin, *Samuel Oldknow and the Arkwrights* (New York: Augustus M. Kelley; 1968, orig. pub. Manchester University Press), 125.
[8] "Pickers book, Nov.–May [1790–1791?]," Samuel Oldknow Papers; R. S. Fitton and A. P. Wadsworth, *The Strutts and the Arkwrights, 1758–1830: A Study of the Early Factory System* (Manchester: Manchester University Press, 1958), 54; John Styles, "Spinners and the Law: Regulating Yarn Standards in the English Worsted Industries, 1550–1800," *Textile History* 44, no. 2 (Nov. 2013): 145–70.

far afield would a man have to ride to collect and distribute to individual households?[9]

While the myth tells us that necessity is the mother of invention, Oldknow's actual experience demonstrates that worlds seemingly ripe for innovation often must wait. Even when new devices appear, it takes time to make them work, as we saw with John Kay's flying shuttle – and many never do. Those that succeed often operate within new arrangements, and getting these operational itself takes work. Existing artifacts and institutions were adapted piece by piece into the emerging order. Even then, the bottleneck that spinning presented to the increased output of those looms equipped with a flying shuttle still waited decades for a resolution. The Society of Arts offered a premium in 1761 for a new machine that one person could use to spin multiple threads at one time, but none seemed forthcoming as the years ticked by.[10] Economic incentives yielded little in technological returns.

Spinning Revolution I

Beginning in the late 1760s, however, new machines for spinning were tried. The first of these was James Hargreaves' spinning jenny, which followed the path suggested by the Society of Arts: it took the work of the Great Wheel, the very large spinning wheel which was itself a medieval innovation, and multiplied the spinner's efforts. Instead of a single spindle, the operator of Hargreaves' machine drew out a beam on which was mounted many spindles instead of one. Where the spinner walked to and from the Great Wheel, drawing out a single yarn, the jenny allowed the spinner to pull out the carriage mounted with many spindles making many yarns. Hargreaves thereby followed the approach suggested by the Society of Arts that had sought six threads spun at one time, but his jenny eventually expanded operations to 130 spindles. (A jenny is a colloquial name for an "engine" – sometimes the myth claims that he named his device after his daughter who tipped over the family spinning wheel and inspired his invention, but no one in his family was named Jenny.)[11] He patented his invention in 1769. By that time he had already tried out his idea, and it had been operational for about five years – for this reason, the

[9] Mary B. Rose, *The Gregs of Quarry Bank Mill: The Rise and Decline of a Family Firm, 1750–1914* (Cambridge and London: Cambridge University Press, 1986), 9, 27–28; Hudson, *Genesis of Industrial Capital*, 61–70.

[10] Christopher Aspin and Stanley D. Chapman, *James Hargreaves and the Spinning Jenny* (Preston: Helmshore Local Historical Society, 1964), 16.

[11] Richard L Hills, *Power in the Industrial Revolution* (Manchester: Manchester University Press, 1970), 56; Christopher Aspin, "New Evidence on James Hargreaves and the Spinning Jenny," *Textile History* 1, no. 1 (1968), 120n1.

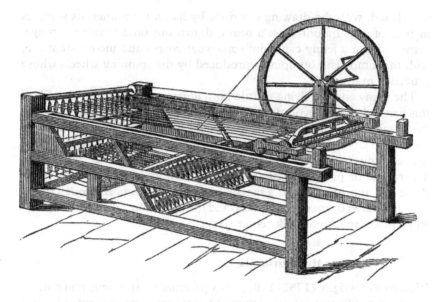

Figure 2.2 James Hargreaves' spinning jenny (ca. 1769) took the traditional work of spinning and multiplied it. Instead of a single spindle to be drawn out by the spinner, it mounted multiple spindles on a carriage that the spinner moved. The pointed tip of the spindle twisted the yarn and the spinner drew it out, an operation similar to that of the spinning wheel. Courtesy of duncan1890/DigitalVision Vectors/ Getty Images.

patent was invalidated almost immediately, in 1770. But spinning jennies remained in use. There is evidence of at least one jenny in Yorkshire still used to spin wool for industry in 1916.[12]

The original 16-spindle jenny slotted well into existing methods and scales of production. Even the larger versions were never driven by water, horse, or steam power – the jenny always operated in places where people worked by hand and used only their own muscles to power the machines. Hargreaves and a partner even built a workshop for jenny spinning. Their Hockley Mill did use horses to run the carding engines that Hargreaves had developed, to better prepare the fiber for spinning. But the jennies themselves operated as the Great Wheel did, and as the domestic spinning

---

[12] Aspin and Chapman, *James Hargreaves,* 23, caption of photo tipped in after p. 60; Stanley Chapman, *The Early Factory Masters: The Transition to the Factory System in the Midlands Textile Industry* (Newton Abbot: David & Charles, 1967), 50.

wheels did, with the drawing out done by hand, for numerous spindles instead of one, mounted on a beam, drawn out on a moving carriage. Yarn spun on a jenny came out somewhat coarse and inconsistent too, with the lumps and bumpiness produced by the spinning wheels whose process it multiplied.[13]

The jenny and the flying shuttle both illustrate the animating spirit of this first Industrial Revolution, the genius of so many machines of the era: they took customary work and duplicated it, multiplying the output of household labor. Hargreaves' spinning jenny mechanized and multiplied the existing techniques of spinning yarn from fiber. It worked in a domestic or proto-industrial setting as well as in the factories, once those had appeared. It was a machine of industrialization, though it was never powered by inanimate forces, and the jenny continued to work even when newer machines took center stage.

Spinning Revolution II

Richard Arkwright (1732–1792) also patented a spinning machine in 1769, on a different model. Arkwright is an excellent example of a real man whose career matches so readily the heroic inventor myth that it was practically made for him. He has been called the inventor of the factory and the father of the Industrial Revolution. He was recognized in his lifetime both locally and as a national figure: he served for a time as High Sheriff of his county of Derbyshire, and his income was the stuff of legend – he was even knighted at age 53. Sir Richard Arkwright was a hero of invention, the sort of great man that would suit the next century. His accomplishments seemed sharper in contrast to his relatively humble origins. He had been a barber for a time, and once owned a pub, but he gave that up to travel the country buying women's hair to make wigs. He had married in his early 20s and named his son after himself, then remarried in 1761 after his first wife died. In 1768, he left Lancashire for Nottingham, where an active knitting industry was beginning to use its established technology, the stocking frames, to knit hose out of yarn spun from cotton. Some say he had already at that point designed the "roller-spinning machine," now more commonly known as the water-frame, for applying waterpower to a frame for spinning.[14]

---

[13] Rose, "Introduction," *Lancashire Cotton Industry*, 10; Chapman, *Early Factory Masters*, 48–49.

[14] Fitton and Wadsworth, *Strutts and the Arkwrights*, 62, quotation at 60; Edward Baines, Jr., *History of the Cotton Manufacture in Great Britain* (London: H. Fisher, R. Fisher, and P. Jackson, 1835), 194; Christopher Aspin, *The Water-Spinners* (Helmshore: Helmshore Local History Society, 2003), 5.

The Arkwright myth ultimately stands on the water-frame, but the device was not the real reason for his accomplishments. First, internalist analysis has led scholars to recognize that Arkwright's water-frame was not quite a new invention. It spun cotton using a method borrowed from throwing silk (a process similar to spinning that twisted and drew out the fiber) that Italians had mechanized, and even powered with waterwheels, since the 1300s.[15] The spindles at the bottom drew the fiber through successive pairs of rollers that narrowed and extended the roving. Each series of rollers moved faster than the last, more finely drawing out the roving until, free of its last roller, it passed through a flyer that twisted it onto the spindle that collected the fresh-spun yarn. Such flyer and spindle arrangements had already long been used on the Saxony Wheel. Using rollers to draw and twist fiber was a well-known approach to spinning. Rollers like Arkwright's had already appeared not just on the Italian silk-throwing machine but had also already been adapted by John Wyatt and Lewis Paul to the spinning of wool and cotton. Paul had patented a machine like Arkwright's thirty years earlier, in 1738. So the idea existed already, but Arkwright made it work. Making the water-frame work, however, required more than machine adjustments.[16]

The water-frame's workings changed over time: it was not a stable machine when Arkwright received his patent for it in 1769. Patents exist in order to protect inventions, and Arkwright used the law to shelter his device even while he was figuring out how to make it work. Applications to patent an invention allowed four months before the specification of its details had to be filed. Arkwright used that time to experiment with the machine and also with the larger scheme within which it ran. For example, while it is usually called a water-frame, named after its power source, Arkwright originally intended it to be powered by horses. Moreover, getting the water-frame to spin yarn also required other devices to prepare the fiber for it. So Arkwright patented a carding machine in 1775, and attempted thereby to link up the whole array of preparation and spinning processes as his own invention. When the time came to file the

[15] Carlotta Bianchi, Fabio Cani, et al., *Guide to the Educational Silk Museum of Como* (Como: Museo didattico della Seta, May 2004), 18; Adam Robert Lucas, "Industrial Milling in the Ancient and Medieval Worlds: A Survey of the Evidence for an Industrial Revolution in Medieval Europe," *Technology and Culture* 46, no. 1 (Jan. 2005), table 2, p. 15; Joel Mokyr, *The Lever of Riches: Technological Creativity and Economic Progress* (New York: Oxford University Press, 1992), 52–68, 103, 212, 232.

[16] Julia De L. Mann, "The Textile Industry: Machinery for Cotton, Flax, Wool, 1760–1850," in Charles Singer, et al., *A History of Technology*, 5 vols. (New York and London: Oxford University Press, 1958), 4:277–78; Walter English, "A Technical Assessment of Lewis Paul's Spinning Machine," *Textile History* 4, no. 1 (1973), 68–70; Timmins, "Technological Change," 34–35.

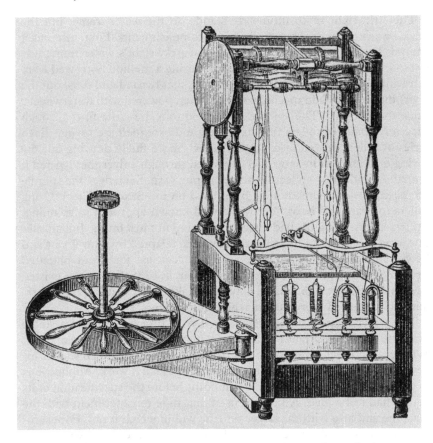

Figure 2.3 Richard Arkwright's water-frame for spinning yarn, patented in 1769, was named for its usual waterpower energy source, though some early examples were driven by horses. As its name indicates, it should be seen in terms of a whole system of production including not just power sources but also raw materials and markets, which were supported by government policies. Courtesy of Nastasic/ DigitalVision Vectors/Getty Images.

specification for the patent on carding, Arkwright asked a friend to draw it up and "make it vague."[17] Being imprecise about his design gave him the rights to gain from technologies that were still shifting as he perfected his

---

[17] Fitton and Wadsworth, *Strutts and the Arkwrights*, 64, quotation at 76; Timmins, "Technological Change," 40–42.

production organization. Bringing it into operation took time, and the idea received government protection while it became an invention.

Arkwright assembled production out of existing components and fiddled with it and with them until they worked together. The individual machines all had precedents: his patents held up just long enough to get his setup running, cement his dignity, and make him a fortune. He had patented his carding engine in 1775. This device used a large spiky cylinder to comb the fibers into one direction, which were then swept off the spikes into a "continuous filmy fleece," which funneled into tall cans as a sliver, the thick web of loose fiber. Next, slivers were drawn together tighter, into a roving, and the roving was put onto bobbins to be fed into a machine to spin it, whether that was Arkwright's water-frame or Hargreaves' jenny. Arkwright's patent for carding was declared invalid in 1781 because other people, including Hargreaves, were already using similar mechanisms. Even his patent for the water-frame expired in 1783, though he managed to hang on to the rights through part of 1785, losing his case in court at midsummer.[18] Nonetheless: linking the preparatory machinery to the water-frame, powering them with a waterwheel and distributing water and power through networks of sluices and gears and shafts and belts, means for feeding and distributing the raw power, and assembling all of these elements in one place, in buildings filled with workers and raw materials, all came together in a technological system now called a factory.

Factory Systems

Sir Richard Arkwright is celebrated for inventing the factory at the same time that the meaning of the word "factory" was changing. Commercial outposts of the East India Company had been called factories after the merchants or factors who worked there. Now a factory was becoming a term to describe a grouping of buildings and devices and energy, to process raw materials or manufacture finished goods – including the transformation of fluffy fiber into yarn to be woven into cloth. Factories worked for many reasons: they organized the knowledge contained in several phases of production into one place, along with the machines and the power sources to drive them along. Linking the machines in a single place also reduced transaction costs: putting-out merchants had divided labor into parts, but they had to physically move their goods from one house to another as value was added and as production progressed.

---

[18] Timmins, "Technological Change," 40–42, quotation at 42; Aspin and Chapman, *James Hargreaves*, 24–25; Aspin, *Water-Spinners*, 80.

Uniting even some of these processes reduced the merchant's costs of doing business. Arkwright's machinery may have been mostly borrowed from other men's minds, but assembling them together into a factory consolidated the production of cloth into the Industrial Revolution.[19]

Factories eventually separated work from home, which meant that people went to work, at prescribed hours, and left when their long days were done. Factories shifted domestic producers into leaving home to work for others, for pay. Time-discipline distinguished factories from artisanal workshops. This means that the pace of work was set by machines, powered by energy sources other than labor's muscle, and this meant workers had to mind them at all times.[20] The abbreviation of the word "factory" from "manufactory" demonstrates the distinction that inanimate power drew between workshop and industrial production. Of course, there were predecessors even for Arkwright's factory organization. In the Midlands county of Derbyshire, efforts to mechanize silk production had already borrowed Italian production practices by 1717. Modeled (as Arkwright's water-frame was) on the Italian pattern that employed falling water to draw and twist silk fiber into yarn, John Lombe's silk mill was a memorable employer of inanimate power in factory production. According to one famous account, its giant water-wheel turned 26,586 individual wheels that powered 97,746 specific movements, thereby spinning 73,726 yards of silk thread – with every rotation of the giant wheel.[21] Visitors marveled to see the Derby silk mill, which for a moment mechanized textile spinning on a grand scale, half a century before industrialization's invention myth.

Technological predecessors of the factory itself came from many sources. Plantations producing sugar had divided labor and routinized the tasks of enslaved workers in scaled-up versions of the household.[22] Plantations cultivated commodities that were consumed far from the fields. Workshops were obvious models for factories, where masters trained apprentices and produced consumer goods, in both guilds and households. Putting-out merchants like Samuel Oldknow distinguished

[19] Philip Lawson, *The East India Company: A History*, 2nd ed. (London and New York: Longman, 1993), 46; Oxford English Dictionary Online, s.v. "factory, n.," accessed 27 May 2017; Lindy Biggs, *The Rational Factory: Architecture, Technology and Work in America's Age of Mass Production* (Baltimore: Johns Hopkins University Press, 2003).

[20] Keith Thomas, et al., "Work and Leisure in Pre-Industrial Society (and Discussion)," *Past and Present* 29, no. 1 (Dec. 1964): 29–66.

[21] William Hutton, "The History of Derby," 191–209, in D. B. Horn and Mary Ransome, eds., *English Historical Documents, Vol. X, 1714–1783* (New York: Oxford University Press, 1969), 458–61; Daniel Defoe, *A Tour Thro' the Whole Island of Great Britain*, 4 vols., 4th ed. (London: S. Birt etc., 1748), III:73–75.

[22] Sidney W. Mintz, *Sweetness and Power: The Place of Sugar in Modern History* (New York: Penguin Books, 1986; orig. pub. 1985), 46–61.

various tasks of production into specific specialist processes, some of which were then mechanized. In some cases, the putting-out merchant owned the machinery and rented it to his workers, while other merchants kept large numbers of hand-operated looms in a single weaving shed. Meanwhile, in the late 1600s, some worsted makers invested capital in building cottages for their workers. They grouped these homes around courtyards, also bordered by warehouses to store raw materials and finished goods. Some merchants incorporated into the complex of buildings the finishing processes of fulling and dyeing and cropping the wool pieces.[23] In these situations, early modern workshops resemble later factories, even if they did not yet link labor processes to power and embody them in machinery.

Other merchant-manufacturers centralized production to keep secrets – not to power machinery but rather to protect some innovation in method or device. One hosier employed more than forty apprentices (the parish paid him £5 each for taking them off the rolls of poor relief, and supplied their clothing during their apprenticeship as well). In the 1720s, he bought a device that attached to his knitting frames and could be used to knit fashionable mitts. He erected a large building in Nottingham, then the home of hosiery- and lace-knitting, to house and hide the workings of his device. Likewise, another Nottingham stocking knitter outfitted his frames with ribbing machines and put them in a windowless room, to conceal their operation. The local populace, employed at similar work and intent on seeing the machinery, climbed on the roof to spy through the skylights.[24] This story adds a different sort of consideration to the development of the factory. While its results were scale of production, power, and centralization, its causes included keeping secrets and controlling access, and its models included the colonial plantations that typified agricultural commodity production.

Richard Arkwright, Heterogeneous Engineer

Arkwright's creation of the factory system is too strong a claim, then, just as his invention of individual machines did not hold up in practice as well as it has served as an invention myth. That should not diminish his accomplishments, however, because his approach incorporated so much more than machines: he also maneuvered laws, workers, and global trade networks into new patterns to make his factories produce a return. His

[23] David Seward, "The Wool Textile Industry 1750–1960," in J. Geraint Jenkins, ed., *The Wool Textile Industry in Great Britain* (London and Boston: Routledge & Kegan Paul, 1972), 43.

[24] Chapman, *Early Factory Masters*, 37–39.

engineering was heterogeneous. As father of the Industrial Revolution, his accomplishments incorporated more than machines – much more than physical stuff. His water-frame drew rovings through the rollers and twisted them through a flyer onto the spindle. The machinery that made the roving would not have worked without the adaption of water-wheels to textile machinery – which themselves would not have worked without abundant waterways and the tradition of using them for milling grain or sawing wood or felting woven wool – a tradition that included the medieval social structure of tenants bound to use the landlord's mill. Existing conditions helped make machines for spinning work. But the keystone accomplishment that made mechanized cotton spinning profitable was a change in the law, and the economic incentives that arose as a result.

It was the second Calico Act that stood in Arkwright's way. Legal machinations around the regulation and therefore the classification of cloth had long played a role in supply and demand. Sumptuary legislation had said who could wear what in the medieval period, and centuries later the gears of government were using laws to alleviate the clothiers' fear of a Calico Craze. The 1701 Calico Act had stimulated the British printing sector by banning imports of Indian printed cloth, but the 1721 law banned even domestically printed all-cotton cloth from sale or use in Britain (though these still found markets abroad). The prohibitions had nursed Lancashire's fustian industry, however, and Parliament taxed its products for revenue: sales along the coasts of the Atlantic provided income to the government. But Arkwright's water-frames spun cotton into yarn coarse but strong enough to use as warp, and this made all-cotton British-made cloth possible. In order to get his water-frame working profitably, Arkwright petitioned Parliament. In 1774, he secured the repeal of the Calico Act from Parliament. Cloth made of "Cotten both ways," both warp and weft, became legal, taxed more than fustian and linen, but lighter in weight and better able to take livelier colors. Consumers liked it, and Arkwright's success in changing the regulatory framework provided incentives for others to adopt his new technological structure. A decade later, tariffs against imported Indian muslins were giving British all-cotton cloth the decisive advantage. Parliament was part of the world that made Arkwright's machine repay its investment.[25]

---

[25] Margaret Cater to Mrs. Mary Williamson, 17 Oct. 1776, quoted in John Styles, *The Dress of the People: Everyday Fashion in Eighteenth-Century England* (New Haven and London: Yale University Press, 2007), 127; Prasannan Parthasarathi, *Why Europe Grew Rich and Asia Did Not: Global Economic Divergence, 1600–1850* (Cambridge and New York: Cambridge University Press, 2011), 111–12, 131.

The Arkwright System

All these elements assembled together made the continuous production of yarn from cotton fiber possible. They made a method of production out of many different technological genealogies not only for spinning, but also for using inanimate waterpower, and for enclosing production within walls that were not homes. In order to operate at a profit, this method needed laws to change. It also used workers and fibers: there will be more on those inputs in a moment. For now, however, let us consider Arkwright's production system as if it were a finished artifact and see how it became one – the elements needed to make it operate, each also within its own context. Basically, Arkwright licensed his ideas to other men, who started spinning mills like his and solidified his methods as well as his reputation. This was an expensive proposition: one firm paid £2,000 to use the design of his water-frame, another £5,000 for the carding engine rights, and another £1,000 per year as a royalty.[26] Licenses secured, the initial investment of actually building the machines and setting up a mill cost about £3,000 more. Licensed operations were "outnumbered by those attempting to work the method unsanctioned," and Arkwright kept up legal actions against such pirates until he finally lost all patent rights to the machinery in summer 1785.[27]

Starting an Arkwright-style mill was neither cheap nor easy. The necessary steps might begin with the purchase or lease of an old grain or sawmill and the erection of buildings and roads. One early adopter discovered that he had to hire men first to catch the moles whose tunnels would undermine his building site. Next came excavating dams and ponds and water-races to turn a large wheel and direct its shafts to run machines; getting belts and raw materials; fixing up carding, drafting, roving, and spinning frames – down to the level of nailing cards onto the frames. It was absorbing work. The man who left the diary describing these steps neglected his farming even after the mill was making Arkwright-style water-twist (coarse yarn, suitable for warp) that needed selling. But the new factory design offered opportunities to fix some of the difficulties men like Samuel Oldknow faced. As a putting-out merchant, his capital resided in the fiber and yarns being worked in other people's hands. His command of the quality and quantity made in those people's homes was limited. He already ran a business that combined elements of industrial with household production, as the workers at home did tasks

---

[26] Chapman, *Early Factory Masters*, 72–74. The value of £1,000 in 1774 would be about £146,000 in 2018.

[27] Gillian Cookson, *The Age of Machinery: Engineering the Industrial Revolution, 1770–1850* (Woodbridge: Boydell Press, 2018), 50; Rose, *Gregs of Quarry Bank*, 20.

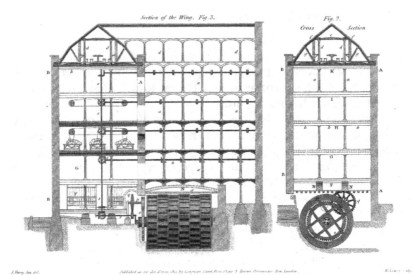

Figure 2.4 An Arkwright-style factory like this one drew power from the waterwheel, seen at bottom. This power was distributed to the factory floors by a system of gears and iron rods, at bottom left – and then carried across each floor by giant turning shafts, labeled "b." Leather belts ran from these shafts to the individual machines; carding engines can be seen on the left, two stories above the shingled face of the wheel. The attic was a schoolroom for the children, likely pauper apprentices, who worked in this early cotton-spinning mill. Courtesy of Photos.com/ PHOTOS.com/Getty Images.

divided more kaleidoscopically than domestic textile production had done. Oldknow took the leap. He was one of the men who paid to adopt the Arkwright model of production.[28]

Samuel Oldknow straddled the gap between putting-out merchant and manufacturer, and his accounts illustrate the transitions that accompanied his adoption of machinery and factory methods. For example, as he industrialized his operations, his workforce changed too, including the gendered division of labor – the different tasks done by men or women. His warps seem to have all been strung by women – and as it happened, a lot of them were named Betty. In 1788–1790, he paid both men and women for spinning – it is likely that men were paid for their family's work. By 1793, as he shifted to factory production, his spinners' book is

---

[28] Unwin, *Samuel Oldknow and the Arkwrights*, 69–84, 162; Aspin, *Water-Spinners*, 85–87.

filled with names of men.[29] Mechanized spinning could be heavy work in
the early days, and by the nineteenth century it had a reputation as skilled
work done by men in the new factories, and Oldknow's records reflect
that switch. Women had been the usual spinners – a spinster was once the
name of an occupation (as a throwster was the name of one who threw silk
fiber), before it became the legal designation of an unmarried woman.[30]
As spinning mechanized, men operated the machines in powered factories, while women still spun at home and were most usually employed at
spinning wheels and jennies. Because these were never powered by inanimate sources, they could still be operated by the fireside, or in small
domestic workshops.[31]

   Another man who invested in the Arkwright plan was Samuel Greg
(1758–1834), whose fortune came from sugar and slaves. His father had
been a merchant and ship-owner who, by 1785, owned plantations in
North America and the Caribbean. The Hydes, his in-laws and business
partners, owned a weaving workshop that seemed always to need yarn. If
Greg could supply it, using the new methods and machinery sold by
Arkwright, his market lay right before him – in the family. Samuel Greg
chose a site primarily for the gushing waterfalls that he would use to power
the machinery. He selected a deep valley, heavily wooded, near a village
called Styal, eleven miles outside Manchester, in Lancashire, in the north
of England. Located on the site of a one-time quarry, he built a brick mill
building in 1784 and called it Quarry Bank. For twelve years, he had
about 150 workers spinning coarse yarn on Arkwright-style water-frames.
Initial growth was slow, but accelerated with the French Revolution,
which added risk and uncertainty but created chances and market opportunity. The British war against revolutionary France started in 1793.
Greg's business grew, as did much of the British cotton industry.[32]

   War with France after 1793, and the Napoleonic battles that thundered
across Continental Europe until 1815, were historical contingencies that
gave British cottons a fighting chance in world markets. Another contributing factor was an English financial revolution after 1688, including

---

[29] "Warpers' Book" (11 Aug. 1804–27 Apr. 1805), first full pages; "Accounts 1788–1792;"
   "Spinners Weekly" accounts for Mar. 9, 1793, in "Spinners' Production Accounts," 2
   Mar. 1793–14 Dec. 1793; all three, Samuel Oldknow Papers.
[30] Oxford English Dictionary, s.vv. "spinster" and "throwster," www.oed.com.lib, accessed
   22 July 2017.
[31] Mary Freifeld, "Technological Change and the 'Self-Acting' Mule," Social History 11,
   no. 3 (Oct. 1986): 319–43; Keith Sugden, "An Occupational Study to Track the Rise of
   Adult Male Mule Spinning in Lancashire and Cheshire, 1777–1813," Textile History 48,
   no. 2 (Nov. 2017), 160–75.
[32] Rose, Gregs of Quarry Bank, 13–20, 46; "Plan of Q B Mill, dam etc., and two fields,"
   [1802], R. Greg and Co. Papers, Manchester Central Library, Manchester, UK.

the invention of impressive financial services and instruments, and the foundation of the Bank of England to fund the government by credit. Neither the Bank of England nor the City merchants financed the industrial sector taking shape in remote Yorkshire and Lancashire, however; "provincial needs were met by local credit networks." Instead the new class of merchant bankers in the City of London lent money through the networks of their personal connections to agriculture and aristocracy. Parliament sat nearby, in Westminster, and throughout our period, the government served mostly landed interests – the 4,000–5,000 lords who owned three-quarters of the land. Landlords in England had benefitted from the enclosure of small farms and medieval holdings into more efficient estates, as well as from grain exports and the same continental disruptions that served the textile sector. After 1815, Corn Laws protected agricultural landlords from competition by levying tariffs on grain imports. Farms contributed the largest share until 1850 of both national income and employment, and paid most of the consumers who bought pocket watches and blue-and-white dishes, and bright fabrics printed with flowers, rather than more raw flax or wool to work up at home.[33]

Even as these specific historical contingencies and structures contributed to the success of the Arkwright order, it still contained surprising technological alternatives. For example: James Hargreaves, inventor of the spinning jenny, has been described as forced into using the Arkwright method. His partner in the Hockley Mill jenny workshop "was compelled to abandon jenny-spinning to become one of Arkwright's earliest licensees." The firm abandoned jenny spinning for spinning cotton warp yarn on an Arkwright-style frame, but they still used horses to drive their carding engines in 1777, which they deployed to make rovings for spinning on the waterframe. Despite these endless adaptations of the technology to specific circumstances, the Arkwright system triumphed: men could buy it and set up shop in the textile spinning industry. By 1788, there were 143 Arkwright-style mills in Britain. By the end of the century, of 900 cotton-spinning factories, fully one-third were Arkwright-style mills with more than 50 workers each. But jenny spinning did not disappear. At the same time that Arkwright-style factories became more common, the jenny persisted. It operated within the muscle-

---

[33] P. J. Cain and A. G Hopkins, *British Imperialism: 1688–2015*, 3rd ed. (Abingdon and New York: Routledge, 2016), 76–79; quotation at 79; P. J. Cain and A. G. Hopkins, "Gentlemanly Capitalism and British Overseas Expansion I: The Old Colonial System, 1688–1850," *Economic History Review* 39, no. 4 (Nov. 1986), 507–08; Styles, *Dress of the People,* 146.

powered domestic setting and also in larger-scale workshops, where apprentices worked under master clothiers, typical of guilds.[34]

Chasing Water

James Hargreaves may have used horses to run the Arkwright system but most licensees used water to power the frames instead. In Hartington, where sixty people worked at home producing yarn and cloth, a man and three sons leased a corn mill in 1790 on the River Manifold and built a factory to spin cotton yarn to have woven into calico pieces. The region had fast-flowing streams and some of the mills built in the 1790s were still spinning yarn well into the twentieth century. Luckily for industrialists, waterfall sites were plentiful in England and had been used to grind grain since the first century CE, when Romans lived in Britain. In 1086, the Domesday book recorded over 6,000 watermills and there were likely more. From at least the late twelfth century, mills were used for fulling and metalworking as well as the grinding of grain into flour. In the cloth-making districts, sometimes more than half the mills were involved in some phase of cloth-making. As we have seen in Chapter 1, the rules for using mills represented a kind of pre-industrial tax, and utilizing its facilities supported the medieval society and its economic structure.[35]

These waterpower sites came in handy for the new factories of the Industrial Revolution. Arkwright's first factory at Cromford in 1771 was powered by a brook. Five years later, in order to expand its operations, he demolished an old corn mill to use its water. Of course, Samuel Greg chose the site for his Quarry Bank Mill at Styal because its watercourses flowed fast, down through steep hills, giving force to his wheels. However, textile machines usually needed to run faster than did the large-diameter waterwheels used to grind grain. While corn mills had gears that trans-formed six or eight revolutions per minute at the wheel to 80–120 revolu-tions per minute at the millstones that ground the grain, even more speed was needed. One solution set gears on the inside surface of the waterwheel so that as it turned, a smaller wheel set inside was spun more quickly. That faster speed was better able to run the Arkwright production system constructed around the water-frame: the smaller wheel turned the great horizontal shafts made of wood, around which were strung leather belts

---

[34] Chapman, *Early Factory Masters*, 51, 92, quotation at 48; Maxine Berg, *The Age of Manufactures: Industry, Innovation, and Work in Britain, 1700–1820* (Oxford: Basil Blackwell, 1985), 190; Rose, "Introduction," *Lancashire Cotton Industry*, 10.

[35] Chapman, *Early Factory Masters*, 58–59; Dickenson, "West Riding Woollen and Worsted," 86–87; Martin Watts, *Watermills* (Princes Risborough, Buckinghamshire: Shire Publications, 2006), 4, 7–8.

that ran to each individual machine and powered carding and spinning engines.[36] Thus millwrights found work in converting existing structures to the demands of a newly emerging industry, and their work produced one part of the setup of mechanized cloth production.

Textile machinery throughout the Industrial Revolution was made by men trained in conventional ways, in the craft apprenticeships that taught them how to work with iron and tin, at furnace and forge. At first the textile machine-making business was "heavily enmeshed with other sectors," from woodworking and carpentry to blacksmiths and ironmongers, but industrial expansion led to increasing specialization. Craft artisans started small firms that clustered around the canal and river networks of Yorkshire and Lancashire. Specialist expertise in textile machinery makes its first appearance in the historical record from the men who testified in the court cases over Arkwright's patents. By 1800, it was rare for textile manufacturers to make their own machines. The insurance policies that covered the textile mills distinguished "clockmaker's work" (the fine metal machinery, in carding engines, roving frames, and spinning frames that Arkwright claimed as his invention) from "millwright's work" (the wheels and gears and shafts that turned the machinery). But the terminology was already inaccurate. Expertise in tin proved far more useful than did the clockmakers' skills in working brass. Nonetheless, the knowledge and skills that both supported and grew from the Industrial Revolution sprang from these older networks and practices.[37]

New machines stood on existing structures. The spinning devices ascribed to James Hargreaves and Richard Arkwright were themselves adapted from earlier innovations made in silk-throwing devices, and the flyers of the Saxony Wheel. They worked best when the raw materials they transformed were also the regularized results of carding and preparatory machinery. It all nestled into older power structures – old mills retrofitted for spinning with new gears and shafts and belts for turning whole suites of machinery. The mills existed, as did arrangements for using streams as power sources. As new or reconceived machinery became operational, it did not entirely displace its predecessors. If one-third of the 900 cotton mills surveyed in 1800 used the Arkwright model, the other 600 were essentially workshops – and many of these used jennies.[38] Arkwright's water-frame did not supersede the earlier jenny. It was not even the signature spinning invention of the Industrial Revolution. That honor

---

[36] Chapman, *Early Factory Masters*, 64–65; Watts, *Watermills*, 46–47.

[37] Cookson, *Age of Machinery*, map frontispiece, 54–56, 69, 85, 107, quotation at 5; Aspin, *Water-Spinners*, 8.

[38] Berg, *Age of Manufactures*, 190.

would have to go to the spinning mule, Samuel Crompton's invention of 1779.

Labor Systems

However, before turning to Crompton's mule, it is time to discuss one more input needed to make Arkwright's invention work. Factories needed operatives, and the workforce that operated jennies and water-frames overlapped. Early factory masters made use of familiar social and economic structures to find factory workers, just as they did with power supplies. The older workshop mode, in which many jennies ran, had precursors in the guilds and the ways masters trained apprentices in their homes. In addition, the eighteenth-century population explosion had resulted in a surplus of children while the Elizabethan Poor Law allowed a church parish to bind pauper children as apprentices; and pauper apprentices, 10–12 years old, filled many of the early textile factories. London parishes did not shrink from sending the poor children in their care hundreds of miles north to work. One 1797 report described the smallest children winding rovings and spreading the fiber for carding while 14–16-year-olds kept mules spinning – their job was to piece together any broken ends of yarn as grown men drew out the heavy carriages and then put them up, returning them to their starting point, laden with spindles and lengths of twisting yarn.[39]

Church of England parishes paid masters for taking on the initial training and upkeep of child apprentices, but some juvenile mill workers did receive wages thereafter. Moreover, housing them and caring for them and getting them to work was no small matter – the adult super-visors were called overlookers – but even with maintenance costs, children were the laborers available to, and chosen by, many early industrialists. The flow of young workers kept many early textile factories afloat – including that of Samuel Greg at Quarry Bank Mill, who found the parish poorhouses a readily "available and renewable" source of poor children to work in his mills, who were "cheaper than the alternatives" (including even their maintenance). In 1790, Greg built them a dwelling and, after 1815, he made an effort to construct a permanent and stable community for all the laborers at Styal.[40] In other cases, erecting a mill enticed workers to locations like Marple, across the river from Oldknow's spin-ning mill in Stockport, which grew from 548 inhabitants in 1754 to more than 2,000 by 1801. Some manufacturers made industrial production

[39] Rose, *Gregs of Quarry Bank*, 28–32; Aspin, *Water-Spinners*, 171–75.
[40] Rose, *Gregs of Quarry Bank*, 28–32, quotations at 20, 30; Aspin, *Water-Spinners*, 175.

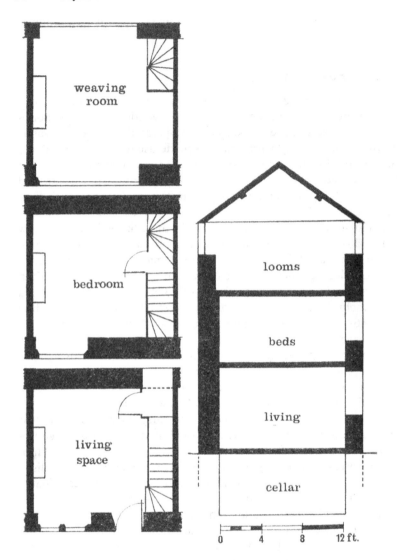

Figure 2.5 Weavers' cottage, 1777, one of many supplied by Richard Arkwright. Valorized as the inventor of the factory and mythologized as the father of the industrial revolution, Arkwright built these sites of domestic production as part of his factory system. Courtesy of the Ancient Monuments Society.

and worker housing part of the same scheme for easing poverty – these philanthropic mill-owners included Robert Owen, who built a model village for spinning cotton yarn in Scotland, at New Lanark.[41] At this point, as the Arkwright system spread, the separation of work from home remained incomplete. Housing apprentices or building villages was one of the costs of doing business.

These children, pauper apprentices who were really factory workers and machine-minders, ask important questions of scholars who claim high wages inspired mechanization and caused industrialization. The influential interpretation of Robert C. Allen rests on this claim. Allen famously compared the wages of construction workers in Asian and European cities and concluded that relatively high wages for builders in London inspired the creation of machines to replace expensive spinsters in northern Britain. Builders and spinsters did not compete for the same jobs so it is hard to see what these data explain. The credibility of this explanation falls further as Allen's data shows that those already-high wages rose more sharply after 1775 than before, and the spike continues until the end of the graph in 1825; if the goal were to lower labor costs, the Industrial Revolution must have failed. In addition, economic historians questioning Allen's hypothesis have demonstrated that earnings were lower than his data indicate, especially in the counties where industriali-zation actually happened.[42] Moreover, through 1790 at least, new spin-ning machines aimed to improve quality and consistency more than the quantities that workers produced.[43]

Historians' arguments about wages, however, provide an opportunity to consider the role of profit incentives and the rationality of economic actors in times of rapid and revolutionary technological change. Even conceding that labor costs were higher in Britain than in India does not explain why mechanized spinning worked because costs change when technology does. A machine that replaces workers (what economists term a capital-labor substitution innovation) will change the type of expense involved in that phase of production, from the operating cost of wages to the fixed sunk cost of capital, which must repay its investment

[41] Unwin, *Samuel Oldknow and the Arkwrights*, 160; Sidney Pollard, "The Factory Village in the Industrial Revolution," *English Historical Review* 79, no. 312 (July 1964), 513.

[42] Robert C. Allen, *The British Industrial Revolution in Global Perspective* (Cambridge and New York: Cambridge University Press, 2009), 34; Ugo Gragnolati, Daniele Moschella, and Emanuele Pugliese, "The Spinning Jenny and the Industrial Revolution: A Reappraisal," *Journal of Economic History* 71, no. 2 (June 2011): 455–60; Robert C. Allen, "The Spinning Jenny: A Fresh Look," *Journal of Economic History* 71, no. 2 (June 2011): 461–64; Morgan Kelly, Joel Mokyr, and Cormac Ó Gráda, "Roots of the Industrial Revolution," UCD Centre for Economic Research Working Paper Series, WP2015/24 (Oct. 2015).

[43] Trevor Griffiths, Philip A. Hunt, and Patrick K. O'Brien, "Inventive Activity in the British Textile Industry, 1700–1800," *Journal of Economic History* 52, no. 4 (Dec. 1992), 893, 895.

over time. Sinking capital in factories and machinery meant a different organization of business from paying multiple workers to add value to raw and semi-finished materials in their homes, at their pleasure. Moreover, the new technology accompanied changes to the workforce – as witnessed by Samuel Oldknow's accounts, where both men and women received pay for domestic outwork but only men spun yarn in the factory. Jobs for men and women paid different rates, and the wages of female spinners were falling in the crucial decades before the 1770s.[44] Finally, new technologies generate unexpected costs: the price of cotton mattered only peripherally to a worsted clothier but was central to the possible returns generated from an investment in Arkwright-style spinning mills.

The people who participated in the Industrial Revolution did not know from the outset that their business would be named any particular thing, nor that its steps specifically resulted in a cotton industry in county Lancashire, in the north of England, organized around machines for spinning fiber grown by enslaved African-Americans in mass quantities on ex-colonial New World plantations. They responded to signals from the whole sphere of business in fabric, from its components and suppliers and equipment to their customers. Their responses flowed from their experiences, the worlds of family and work and food and clothes in which they lived. The prices and valuations of inputs and outputs changed as the technology changed, and while business men were certainly swayed by these economic considerations, their gyrations took shape within the structures and using the steps that they knew and understood. They did not simply want to make more goods more cheaply; they also needed to meet customer demand for specific fabric characteristics, using available sources of labor, fiber, and machinery, even as all these were changing.

The question of pre-industrial wages represents one debate among economic historians of the Industrial Revolution, while the sources of labor present another. The wave of enclosures that modernized aristocratic agriculture between 1760 and 1840 combined narrow cultivation strips into fields and fenced what had been common lands or waste lands into private property. Marxist historians have argued that enclosure severed peasants from the land and from their ancestral relationships with feudal landowners, and farmers without land became factory workers.[45]

[44] Jane Humphries and Benjamin Schneider, "Spinning the Industrial Revolution," *Economic History Review* 72, no. 1 (Jan. 2019): 126–55; Jane Humphries and Jacob Weisdorf, "The Wages of Women in England, 1260–1850," *Journal of Economic History* 75, no. 2 (June 2015): 405–47, Cookson, *Age of Machinery*, 27–28.

[45] Karl Marx, *Das Kapital* (Hamburg: Verlag von Otto Meissner, 1867–1894), vol. 1, chapter 27, entitled "The Expropriation of the Agricultural Population;"

Other scholars have claimed that increasing agricultural productivity, reaching revolutionary levels, provided not only spare labor, better nourished for work, but also the spending power that created consumer demand, stimulating the market for new imports and for the goods emanating from workshops and factories.[46] Even with mill villages and modernizing agriculture, however, textile mechanization only created fresh demand for children who had always worked, on farms, as apprentices, or for crumbs. Yet the new industrial order in fact developed and relied upon tasks specifically suited to child workers. The technological and labor systems developed hand in glove, and one fit the other very well as a result. In addition, one generation that went out to work when very young usually produced another, and child labor became part of the textile industry for the next forty years, and beyond – until nineteenth-century legislation began to set limits on age of employment and hours of work.[47]

These constellations of labor forces, power sources, sites of production, buildings, machinery, merchants and markets, patriarchal family and social class structures, property rights and Parliament's laws, still represent only a selection of the elements that were assembling to become the Industrial Revolution – its several versions. The trade in Indian cotton cloth and the quasi-governmental institution, the joint-stock East India Company, also played a role, as did the cotton plantations of the American South, the Atlantic slave trade, and African laborers, merchants, and consumers. Before fleshing out the framework just a little more thoroughly with these details, however, there is one more spinning machine to be added to the list of inventions. This is the device most closely associated with the Industrial Revolution, with the cotton industry of Lancashire, and its mechanization in the late eighteenth century. This device for spinning yarn was Crompton's mule.

Samuel Crompton and the Spinning Mule

Samuel Crompton (1753–1827) was from Bolton, a village ten miles from Manchester with a deep history in the textile industry of Lancashire. His parents were described a century later as classic domestic

E. P. Thompson, *The Making of the English Working Class* (New York: Vintage, 1966; orig. pub. London: Victor Gollancz, 1964), 214–33.

[46] Mark Overton, *Agricultural Revolution in England: The Transformation of the Agrarian Economy 1500–1850* (Cambridge: Cambridge University Press, 1996); Jan de Vries, *The Industrious Revolution: Consumer Behavior and the Household Economy, 1650 to the Present* (Cambridge and New York: Cambridge University Press, 2008).

[47] Jane Humphries, *Childhood and Child Labour in the British Industrial Revolution* (Cambridge and New York: Cambridge University Press, 2010), 250.

producers on the family labor model – they were tenant farmers who carded, spun, and wove cloth. Their goods likely sold on market Mondays to the merchants from Manchester and London who came to Bolton for its characteristic product, a heavy fustian that wove fine but cheap cotton into a warp of linen yarns imported from Belfast. Like many children in the textile regions, Crompton spun yarn as a child. At age ten he was taught to weave and thus began his apprenticeship in the more manly occupation, the job of the household patriarch.[48]

As a weaver, Crompton found himself "grieved at the Bad Yarn" he had to weave.[49] He did not understand why "if I could find one good thread of a yard long there was ten bad ones and being grieved at this I reasoned thus in my own mind if one thread can be good why not all good?" This was the problem he set out to solve. His house, called the Hall-i'-th-Wood, was part of a rambling mansion outside Bolton, with dark spare rooms that could accommodate his tests. He spent about three years on "very many reapeted [sic] trials and Experiments" before he settled on an approach to the problem, then made a "maschine which was to be the Standard." He had done some spinning on a Hargreaves-style jenny and understood its basic operation. One thing he learned from his efforts was that continuous operation made better thread – that stopping and starting the mechanism made bumpy, inconsistent, yarn, as the jenny did. By duplicating the hand of the spinner with a moving carriage and thereby spinning several threads at one time, the heart of the jenny's innovation, Hargreaves' jenny made a homespun thread, thin and weak in some places, thick and lumpy in others.[50]

Crompton's finished machine kept the jenny's moving carriage, and mounted the bobbins upon it, so that the twisting of spun yarn was done while drawing out the carriage and the yarn. As the carriage was put up, or returned to its starting point, the spinning bobbins gathered up the twisted yarn. In Crompton's mule, the spinning wheel on which Hargreaves' jenny had relied was replaced with rollers, as in Richard Arkwright's water-frame, so the roving had already been made finer and tighter by rollers when the spinner started drawing out the carriage. These rollers tightened the roving up before drawing it out and twisting it into yarn. When operated continuously, this device made a smoother and

---

[48] Gilbert J. French, *The Life and Times of Samuel Crompton*, 2nd. ed. (Manchester and London: Thomas Dinham & Co. and Simpkin, Marshall & Co., 1860), 6–9, 26; Crompton to Sir Joseph Banks, draft, Oct. 30, 1807, Samuel Crompton Papers, Bolton Archives and Local Studies Service, Bolton, UK.

[49] Samuel Crompton to Gentlemen, Dec. 30, 1802, Crompton Papers.

[50] Crompton to Banks, Oct. 30, 1807; French, *Life and Times of Samuel Crompton*, 28, 35; Chapman, *Early Factory Masters*, 48–49.

more consistent yarn from each of its spindles. The trade immediately noticed.[51]

Everyone wanted to see Crompton's "maschine," so much that the requests interrupted his own business. Manchester's manufacturers subscribed money for the opportunity to see his apparatus; he received about £400 from them, and the ideas embodied in his device began to spread across the region and across the textile industry. Later, when he tried to establish his own claim to the invention, he wrote the Royal Society in 1807 and directed observers to see the operations of many of the most important spinning firms, from Horrocks of Preston to McConnell & Kennedy of Manchester "or any Respectable House in Glasgo" to support his claim of priority in invention.[52] He probably should have directed these accounts to the Society of Arts rather than the Royal Society, as the former was the institution that had offered a premium for spinning improvements back in 1761. Its members also sometimes scrutinized unpatented inventions and specifications in order to resolve priority. His confusion about the London scene probably sent his letter astray.[53]

Crompton was never really rewarded for his invention. His detailed descriptions of his device and its development were efforts he made later in life, to express what he had done, in hopes of making some money from the wide employment of his mule. In the common version of this story, Crompton was foiled by institutions. Because his spinning machine was a mule, a combination of existing patented machines, it was not a real invention, patentable under British law. But while this depiction is true, it also functions as a technological myth, a way of sweeping under the rug the vast complexity characteristic of the history of technology. For one thing, British patent law was hardly a rational institution designed to foster invention or to protect inventors. Instead it was a holdover from a more antique way of doing business. Certainly patenting was expensive, and Crompton may also have been ignorant of its operations: Christine MacLeod, great historian of British invention, says: "Patenting seems never to have entered his head, until it was too late." As Hargreaves also used and distributed his machine before patenting it, this seems to have

---

[51] Abbott Payson Usher, *A History of Mechanical Inventions* (McGraw-Hill, 1929), 266–67.

[52] Samuel Crompton to the Merchants, Manufacturers, Spiners [*], Printers & c engaged in the Cotton Trade in These United Kingdoms, Apr. 22, 1811, Crompton Papers; Crompton to Banks, Oct. 30, 1807.

[53] French, *Life and Times of Samuel Crompton*, 142–43; for the role of the Society of Arts, see Fitton and Wadsworth, *Strutts and the Arkwrights*, 39–40; MacLeod, *Inventing the Industrial Revolution*, 194–95.

been a common difficulty for the men connected with the inventions of the Industrial Revolution.[54] Even Arkwright's successful enforcement of his patent rights, under which he licensed others to use his machinery, faltered after his initial exploits.

Crompton's brain-child had already slipped out of his control. By 1811, the mule had been widely employed in textile spinning, and still Crompton pined for recognition and for money. He conducted what he called a "spindle enquiry" in the sixty miles around Bolton. In just that small region of the world, he found 650 cotton mills, and then counted the individual spindles on all the spinning machines they used, to compare which machines were being used more. He counted spindles on Arkwright-style water-frames and on spinning jennies, believing that his invention would demonstrate its superiority by comparison. There were about 156,000 jenny spindles in operation and 310,000 spindles on water-frames – compared to more than 4.2 million spindles on mules that had been modeled on his. The numbers were likely even higher: although the mule was by 1811 prevalent in spinning woolen yarns as well as cotton, he counted spindles only in cotton mills.[55]

Crompton was an archetypical manufacturer-inventor, and his failure to earn the monetary rewards of his great invention lies partly in this personal characteristic. He invented for himself and to improve his own work. That others importuned him to see what he had done only demonstrates the efficacy of his efforts. He did what those in his position often do, and his success at meeting his goal meant his failure in the larger world that was changing from merchant to industrial capitalism. He died penniless, composing hymns for the Swedenborgian church, while his machine made fortunes for some of the many men who succeeded as production moved from home to factory, powered partly by Crompton's mule. Not until thirty years after his death would his role as a hero of invention be recognized and celebrated. Beginning in 1853, the centenary of Crompton's birth, a Bolton cloth merchant worked to erect some public honor for the man whose mule spun the region's industrial wealth. Town leaders and cotton manufacturers hesitated and delayed, while working men and women teemed the meetings, bursting with "personal pride in the inventor's ingenuity and a fierce sense of possession." They subscribed the funds for a monument "to remind the employers that their wealth rested on the skill and ingenuity of their workers." The

---

[54] MacLeod, *Inventing the Industrial Revolution*, 76, quotation at 93; Aspin and Chapman, *James Hargreaves*, 23.
[55] "Spindle Enquiry Papers," 1811, Reel N, Crompton Papers.

1862 statue of Samuel Crompton, relaxed in his chair, served the purpose, and elevated a real man into a myth.[56]

Side by Side

Technologies work when they fit into their worlds, while both are changing and accommodating and solidifying the other. Different techniques and devices for spinning cotton existed side by side for a number of reasons, and these incentives reveal the genius of the age, of the early Industrial Revolution: its emphasis on quality and characteristics over cost competition. The qualities of the finished product were just as important as its price in deciding whether or not the machine that made it had a chance of running profitably. Costs mattered, but only within categories which themselves were changing as fashionable fabrics changed, and included clothiers' and consumers' calculations about whether one yarn or cloth might satisfactorily substitute for another. Some cotton mills ran both Arkwright-style water-frames and spinning jennies, because each spun a different yarn. Arkwright's water-twist worked for warp, but it was too coarse for most weft. In a further differentiation, Crompton-style mules spun yarn in one direction, creating a z-shaped twist, but a jenny turned the yarn in the other direction, imparting an s-twist. The jenny's operation resulted in an irregular, fluffy, lofty yarn that was excellent weft but not strong enough for warp. These two yarn types meshed well: fabrics woven from mule-spun warp and jenny-spun weft were smooth and absorbed dye well – an optimal ground for printing.[57] Both yarn types were needed for the best all-cotton cloth, which Arkwright had successfully urged Parliament to decriminalize.

Economic historians who study rates of invention and patents during the Industrial Revolution have concluded that most inventions of the 1760s, 70s, and 80s, focused on increasing the types and finishes of cloth produced rather than saving inputs and lowering production costs. Product innovation rather than cost competition characterized the early phases of textile industrialization. Efficiency was not the main goal of mechanization, even though it was one of the results. In fact, the word "efficient" did not yet even describe economic considerations. It was first a philosophical concept that described "agents and causes of change," as the efficient branch of government executes the designs of the whole, or as the efficient cause of dew is the cold temperature of night, warming with

---

[56] MacLeod, *Inventing the Industrial Revolution*, 93; MacLeod, *Heroes of Invention*, 297–303, quotations at 301, photograph at 302.

[57] Berg, *Age of Manufactures*, 175; Philip Sykas to Barbara Hahn, 9 March 2016, email in the possession of the author.

dawn.[58] This concept was transformed, like so much else, into "an industrial invention, created by engineers and physicists to measure the performance of machines, and, in particular, to relate a machine's output to the inputs it had used." This changing definition illustrates how quality, more than quantity, was the goal for which these men were reaching. Crompton was sorely grieved by bad yarn, and European textile producers were learning how to make cotton cloth that reached Indian "specifications and standards."[59] A wide array of technological means made an even larger number of cloth types, and not all were entirely made of cotton. Other fibers, too, played an ongoing role in the Industrial Revolution.

Familiar Fibers

The cotton industry of Lancashire was leaping off the shoulders of the existing British textile business, which was also undergoing a version of industrialization, analogous to the classic case of cotton. The operational definition of the process (mechanization, the separation of work from home and consumption from production, and the regularization of both distribution and the getting and working of raw materials) also measures more common fibers. Woolen and worsted production also grew behind the wall of protection erected by the Calico Acts. In Yorkshire, wool stayed small: a household made a piece or two per week and sold it at public markets, like the one in Leeds, or peddled it to a larger producer who wholesaled it with his own goods. Worsted had been proto-industrial in the region from the start, one of the New Draperies sold by the heavily capitalized merchants around Halifax. Such men might buy fleece at the biggest fairs and put it out to separate households for combing, spinning, and weaving. In other places, worsted was made at home on a small scale, so the capitalistic Yorkshire worsted merchant was not the only way to organize the business: there was nothing in the fiber or processes that demanded the larger investment. Merchants selling worsted cloth pieces in West Africa or North America sometimes placed orders for fabric made to their specifications, while in slower times they could turn to the local markets and Cloth Halls to meet the

---

[58] Oxford English Dictionary, s.v. "efficient," accessed 25 June 2017; Griffiths, Hunt, and O'Brien, "Inventive Activity," 895.

[59] Jennifer Karns Alexander, *The Mantra of Efficiency: From Waterwheel to Social Control* (Baltimore: Johns Hopkins University Press, 2008), 16–26, quotation at 2; Parthasarathi, *Why Europe Grew Rich*, quotation at 90.

demand. The flexibility of the markets, and of demand, made household production persist.[60]

But Yorkshire merchants were selling directly to North American dealers, bypassing London and gaining access to newer information about
what consumers wanted. Attending to the vagaries of distant fashion
called for new products. A least some merchants expanded into large-
scale workshops for finishing their purchases; it was there that forbidden
gig-mills were employed, threatening the independent croppers who
finished the wool. Increasing scale of domestic operations and the rise
of large-scale workshops, the use of machines for some technical processes and the gathering of work into locations devoted to production: all
these existed at the same time, and "depended on the same infrastructure
of trading links, fulling mills, and even labor." One merchant could both
direct a cloth-making workshop and also buy pieces from their makers in
the well-worn way. So familiar fibers had alternate versions of industrialization, which shared some features but also showed important differences to the cotton industry of Lancashire. Spinning was the mechanizing
heart of the business, and a mill-owner might shift the fibers he spun when
technology was changing. By the end of 1792, there were about 100
cotton mills in Lancashire and another 100 in Yorkshire, with about 80
woolen or worsted mills, and some of the cotton firms shifted into worsted
spinning, or vice versa, as the century turned.[61]

Linen (made from flax plants) provides yet another model of industrialization in eighteenth-century Britain. Fustian-makers had used linen
warps imported from Ireland and the Continent, and in the eighteenth
century, British merchants turned manufacturers by mechanizing some
phases of flax production. The hero of linen industrialization was John
Marshall (1765–1845) of Leeds. Marshall was a merchant's son and had
proper training in foreign markets as well as the domestic production side
of the business. When he was 22 years old, in 1787, his father died
suddenly, leaving him in complete command of a linen firm. Only a few
weeks later Marshall set out to investigate reports of a flax-spinning
machine. He went to Darlington, just beyond Yorkshire's northern
boundary, and an important center of British linen manufacture. Here,
a former worsted weaver, working with a watchmaker, funded by a linen
manufacturer, had built a flax-spinning frame on the principle of one he

[60] Berg, *Age of Manufactures*, 209, 215–17; John Smail, *Merchants, Markets and Manufacture: The English Wool Textile Industry in the Eighteenth Century* (Basingstoke and New York: Macmillan and St. Martin's Press, 1999), 68–70.
[61] John Smail, *The Origins of Middle-Class Culture: Halifax, Yorkshire, 1660–1780* (Ithaca and London: Cornell University Press, 1994), 54–55, quotation at 54; Hudson, *Genesis of Industrial Capital*, 156; Cookson, *Age of Machinery*, 44.

had seen at work in Lancashire. When patented in 1787, this was registered as a device to spin wool yarns as well as flax and hemp. This machine
was never very satisfactory, but it found licensees, men who would pay to
use the new machinery – including John Marshall. License in hand,
Marshall took the plans for a spinning frame from Darlington back to
Leeds.[62]

There, Marshall focused on carding and spinning but achieved little
with the machines he had paid to use. By the end of 1788, he hired
Matthew Murray (1765–1826) to help. Murray had artisanal training as
a whitesmith, someone who fashioned articles out of tin. Like many
people who attempted to mechanize spinning, he came to focus on the
preparatory devices to help make the spinning machines work. As with
any fiber, flax must first be turned into a kind of semi-yarn before spinning. Retting, scotching, and heckling worked to rot the flax in water,
remove its straw-like exterior and break down its woody core, and comb
and straighten the remaining fibers to lie in parallel. The next step,
intended to draw the heckled fibers increasingly fine, would create
a fibrous web (like a sliver or roving) which spinning would strengthen
by drawing out and twisting for strength. Marshall and Murray experimented extensively with heckling and eventually registered a patent,
bought a site on the south side of the River Aire, and equipped it with
flax machinery. By 1793, they had a Boulton and Watt steam engine in
operation.[63]

Marshall's story shows that the textile business of spinning yarn
changed technologically for many fibers, just as for cottons, in the late
eighteenth century. Many fibers continued to be spun, and even more
rovings, yarns, and cloth types were made from them. Fibers other than
cotton attempted and experienced the same sort of mechanization –
multiplication of worker effort – as did the spinning machines tied to
Richard Arkwright's reputation. Thus we find that John Kay's flying
shuttle was only one yarn, one story among many, within a wider context. The flying shuttle was part of a system for opening and dressing
wool, the Arkwright water-frame was part of a complete organization for
cotton yarn production, and the linen industry that mechanized in the
generation after Arkwright also worked within a licensing model for

[62] W. G. Rimmer, *Marshalls of Leeds: Flax-Spinners, 1788–1886* (Cambridge: Cambridge University Press, 1960), 13–24; Gillian Cookson, *Victoria County History: A History of the County of Durham* (Woodbridge: Boydell and Brewer for the Institute of Historical Research, 2005), 4:154–56.

[63] British Patents #1752 (1790) and #1971 (1793); Rimmer, *Marshalls of Leeds*, 26–27, 29–38; John Marshall, "Notebooks on Experiments on Spinning," July 1790–June 1801 and Aug. 1803–1823, Records of Marshalls of Leeds, Special Collections, Brotherton Library, University of Leeds, UK.

business and mechanization. But all these technological assemblages that we have described as larger than machines were bigger than even these descriptions have conveyed. To see the bigger picture, local history provides an appropriate lens. Now it is time to examine the neighborhood contingencies that shaped the adoption and working of the cotton spinning machines, and at the same time to carry the tale of the Industrial Revolution well beyond any particular town, county, or country. Now it is time to turn to Manchester.

Suggested Readings

Aspin, Christopher. *The Water-Spinners*. Helmshore: Helmshore Local History Society, 2003.

Berg, Maxine. *The Age of Manufactures: Industry, Innovation, and Work in Britain, 1700–1820*. Oxford: Basil Blackwell, in association with Fontana, 1985.

Chapman, Stanley. *The Early Factory Masters: The Transition to the Factory System in the Midlands Textile Industry*. Newton Abbot: David & Charles, 1967.

Hudson, Pat. *The Genesis of Industrial Capital: A Study of the West Riding Wool Textile Industry, c. 1750–1850*. Cambridge: Cambridge University Press, 1986.

Rose, Mary B., ed. *The Lancashire Cotton Industry: A History Since 1700*. Preston: Lancashire County Books, 1996.

Styles, John. *The Dress of the People: Everyday Fashion in Eighteenth-Century England*. New Haven and London: Yale University Press, 2007.

Unwin, George. *Samuel Oldknow and the Arkwrights: The Industrial Revolution at Stockport and Marple*. Manchester: Manchester University Press, 1968.

3   Cottonopolis

The invention myth goes beyond machines. The term "Industrial Revolution" usually refers not only to individual devices but to the whole cotton industry of Lancashire, and to its best-known site, Manchester. In the 1770s, chasing water had led to the proliferation of water-driven Arkwright mills on sites of medieval industry, organized by feudal relationships. As the 1770s ended, industrialization urbanized. Manchester – the world's first industrial landscape, with its monumental, repetitive, red brick mills, and its spewing smokestacks and filthy tenements – was coming into being. It was a real, physical, material place and it was also a myth that confers order on a bigger story – just as with the spinning devices, the machines of the Industrial Revolution. Probing behind the myth again offers a more thorough story of industrialization, because local choices and contingencies shaped the path of technological change. The men who ventured from Manchester to Bolton chose Samuel Crompton's spinning mule, and this choice contributed to the mass production of yarn, as had Richard Arkwright's Parliamentary maneuverings. But the local was global, and decisions made in Manchester had far-reaching causes and effects. From India to Africa and the Americas, Manchester's expanding manufacturing sector was enmeshed in world trade. But all the shifting markets and structures of empire, changing systems of work and production, and reverberations of taste and design and consumption, found their expression alongside the black greasy waters of the Ashton and Rochdale canals, threading their way through Manchester's Ancoats neighborhood. Looking closely at the local demonstrates how Manchester's links spanned the globe.

Manchester had been a textile town for hundreds of years. Conveniently situated at a confluence of several rivers, where the Irwell met the Medlock and the Irk, the mounded riverbanks nursed markets for the fabrics made across southeastern Lancashire. The place was not yet a city nor even a town, and some say that the lack of craft guilds and a formal town corporation aided Manchester's rise. The place was a manorial borough governed by a Court Leet, a medieval form of governance set up in Manchester back in 1301. Its textile business adapted well to the early modern world: the coarse wool cloth from the region sold comfortably under the name "Manchester cottons" in the late 1500s, and as the century turned, its makers took up fustian production. By 1750, the Manchester region made 114 types of cloths, its various mixtures woven from cotton, linen, silk, and woolen fibers. A distribution center for its environs, in 1773 the town included 159 textile tradesmen who based their operations in the surrounding villages and towns but had their warehouses in Manchester, where they met suppliers and customers. More than two-thirds of these specialized in fustians, and woolens dealers made 15 percent of the total.[1]

Manchester's influence sprawled over its surrounding towns and villages. In the 1770s, these included fustian towns Bolton and Oldham, where weavers received linen warp from Ireland, Germany, and the Netherlands, and cotton from the Levant. Of the several ways to organize the textile business, many were in operation in the Manchester hinterlands in the late eighteenth century. Some spinners and weavers worked in their homes for the merchants; some were independent operators. Some owned their looms, and others had machinery supplied by those who distributed their finished goods. The bleaching, printing, and dyeing industries had also flourished along the rivers of Lancashire, under the protection of the eighteenth-century Calico Acts, and this innovative industry deepened the region's commitment to the textile business.[2] So Manchester was the market for a wide range of textiles, and an array of production systems for making them, each with its typical distribution outlets and raw material suppliers.

New transportation links spread Manchester's trade connections in the eighteenth century. At the start of the 1700s, roads and bridle paths linked the town to the surrounding country. Turnpikes authorized in the 1720s and 1750s connected Manchester to communities including Oldham and Bolton, Stockport, and east into the Yorkshire networks

[1] Alan Kidd and Terry Wyke, *Manchester: Making the Modern City* (Liverpool: Liverpool University Press, 2016), 5–11; Geoffrey Timmins, "Roots of Industrial Revolution," in that volume, 43, 56–57; Alfred P. Wadsworth and Julia De Lacy Mann, *The Cotton Trade and Industrial Lancashire, 1600–1780* (Manchester: Manchester University Press, 1931), 13–15; Trevor Griffiths, Philip A. Hunt, and Patrick K. O'Brien, "Inventive Activity in the British Textile Industry, 1700–1800," *Journal of Economic History* 52, no. 4 (Dec. 1992), 894.

[2] Peter Maw, *Transport and the Industrial City: Manchester and the Canal Age, 1750–1850* (Manchester and New York: Manchester University Press, 2013), 5, 7; Geoffrey Turnbull, *A History of the Calico Printing Industry of Great Britain*, ed. John G. Turnbull (Altrincham: John Sherratt and Son, 1951), 19–25, 99–109.

including Leeds and Halifax. These turnpikes also ran all the way west to Liverpool, Lancashire's port to the Atlantic, on the banks of the Mersey River. The 1721 Mersey and Irwell Navigation Act had canalized a route between Manchester and Liverpool, and, by 1772, there were twenty-one vessels supplying regular carrying service. The 1765 Bridgewater Canal linked Manchester to the Duke of Bridgewater's coal mines and provided a route to Liverpool as well. In the 1790s, the city planned many more water networks, including the Rochdale Canal and the Ashton Canal, with impressive infrastructure – wharves, warehouses, turning basins. Canal systems served the commerce that linked Manchester's surrounding communities to more distant traders too, including men who carried Manchester goods to Europe and the Americas. The town's finishing facilities readied regional cloth for its markets. It was well situated to handle the output of mechanized, power-driven production as manufacturing moved from homes to factories in the following decades.[3]

Exploring Manchester during industrialization reveals how the new technology came out of what already existed. We know the commercial basis of the emerging manufacturing sector, and have witnessed merchant capitalists shifting the patterns of world trade. We saw Samuel Oldknow transition from putting-out merchant to manufacturer by licensing and building an Arkwright-style mill at Stockport. Samuel Greg also relied on existing institutions when he used the apprentice system to make pauper children into machine minders at Quarry Bank. Some of his money, too, came from enslaved labor on Caribbean plantations, which are generally associated more with merchant than industrial capitalism.[4] But during industrialization – almost by definition – manufacturers parted ways from the commercial side of the business. As manufacturing took shape, its industrialists began to get their cotton through brokers in Liverpool. And Liverpool's history, from the slave trade to the cotton boom, will sketch the way that England's networks encircled the globe, from the American plantations that grew the cotton to the expanding commercial empire in India.

Manchester Manufacturers

To uncover the links of Manchester, let us turn to one of its most famous cotton firms. McConnell and Kennedy were both Scotsmen, from the village of Kirkcudbrightshire, and both had served apprenticeships to

---

[3] Maw, *Transport and the Industrial City*, 5, 7, 35, 71–74, 238–39; Timmins, "Roots of Industrial Revolution," 59, 76–78.

[4] Mary B. Rose, *The Gregs of Quarry Bank Mill: The Rise and Decline of a Family Firm, 1750–1914* (Cambridge and London: Cambridge University Press, 1986), 25.

machine makers. Their story starts when James McConnell made two spinning mules for a customer, who failed to pay; left with his investment in the machines, he decided to set up and operate them himself. By 1791, he was working with John Kennedy and, within four years, the pair were partners in a venture that spun yarn on mules and made mules to sell to other spinners, and that supplied rovings to other people to spin into yarn on mules. Their business model recalls some important points about this early period of industrial capitalism and its emphasis on quality rather than price. McConnell and Kennedy were not in competition with other spinners to make the lowest-priced goods. Instead most firms specialized in yarns of particular quality and characteristics, which meant they could buy suitable rovings and focus on spinning to specification. Men who made mules were not trying to stop others from spinning, even though they spun too. They could sell their customers rovings as well as mules and profit from the preparatory processes. As new technological systems took shape, this was not yet quite an era of cost competition.[5]

The firm's day-to-day operations also add to our grasp of the fitful, contingent nature of technological change. Their shipping arrangements were characteristic of preindustrial commerce and it is worth remembering just how goods moved along canals and turnpikes: "Your Four Mules will be ready for Loading and packing on Saturday ... it may be best to Load the wagon first with what it can take and then pack the remainder In Box, to go by water you perhaps can send some Boxes by the Wagon."[6] Transportation links did not make commerce entirely smooth. Some assembly was required. In the early phases of industrialization, too, the buyers had to learn how to use the machines, and so did their workers. McConnell & Kennedy did "Hope the Engine Wheels will remedy the Number of soft ends which is met by making the rovings all one thickness" as variations "may be one cause of so many little snarles and a little advice to the reelers to take out the pieces of waste and some loose Ends" might help the machine work better.[7] Telling reelers how to work or changing the engine wheels: either might help make the machines function as their investors wished.

Customers also regularly suggested design modifications or tested new ways to arrange the parts and the processes. When one customer discovered that tin rollers made the "Mules go so heavy," McConnell &

[5] Finding aid, McConnell & Kennedy Papers, John Rylands Library, University of Manchester, UK; Griffiths, Hunt, and O'Brien, "Inventive Activity," 893.
[6] McConnel & Kennedy to Taylor & Haywood, 15 Dec. 1795, Letter Book 11 Feb. 1795–31 May 1796, pp. 160–61, McConnel & Kennedy Papers.
[7] McConnell & Kennedy to Taylor & Haywood, 6 July 1795, Letter Book 11 Feb. 1795–31 May 1796, pp. 62–64.

Kennedy suggested replacing the solid rollers with "Drums" as an alternative, while some who had bought rollers later exchanged them "with the
man that makes the Drums"; possibly to melt one down to make the
other. The mule-makers had to interpret the needs of customers and
convey those wishes to the firms that made some of the parts to their
mules. To the firm that supplied the rollers, they wrote, "When we
received your rollers we did not Observe that the holes were so far from
the Center of the Roller, owing to the Spindle you turn them on not being
true, we therefore wish you to remedy that before you make the next for
us."[8] Only practice indicated what machines and processes worked, but
drilling holes off center probably made machines run poorly.

Getting machines working meant sending along men, erectors and
engineers, "despatched to help 'bed in' equipment" to existing structures. These hands-on installations, the extended invention process,
the negotiations between the machine-makers and their customers, all
undermine any lingering sense that the technology was the product of
a flash of discovery on the part of an individual inventor. The technological system was under construction as machines were made to
run. These letters about drums and rollers, soft ends and snarls, were
sent in 1795, decades after Arkwright's system was stable enough to
transfer to new locations, and fairly late in the adoption of mule
spinning, the mechanization of production, and the development of
the factory system. Yet the correspondence demonstrates that the
operation and composition of the technology – even of the machines
themselves – was still evolving and interacting with other parts of the
system. Getting mules going and mechanizing spinning employed
canals and roads and shippers, the rovings and preparatory machinery, the operators of mules and the suppliers of parts. Reelers – people
who operated preparatory machinery – had to be incorporated into
the system in order to make the spinning machines run.[9]

Factory buildings also formed part of the technological system, and
their changing construction also figured in choices about machines. The
multistory, red-brick buildings, made of repeating, identical bays, each
with a large window to admit light, had become the recognizable standard
mill. Like the machines themselves, the factories permitted multiplication
of work. And the number of bays proliferated, spurred by the duplication
of spindles on single machines, which then made smaller buildings

---

[8] McConnel & Kennedy to Taylor & Haywood, 16 May 1795, 11 Nov. 1795; idem to
Thos. Harrison, 26 June 1795; all three, Letter Book 11 Feb. 1795–31 May 1796, pp. 58,
62–63, 135; "Time Book," (n.d.), Samuel Oldknow Papers, John Rylands Library.
[9] Gillian Cookson, *The Age of Machinery: Engineering the Industrial Revolution, 1770–1850*
(Woodbridge: Boydell Press, 2018), 228.

inoperable.[10] By February 1795, standard spinning mules were too big for small factories. McConnell & Kennedy recommended that customers should buy as large a machine "as their Room admit it." The firm no longer made mules of 144 spindles, 180 was "as few as any we have made this some time" and while 144 was fine for hand spinning, the gearing to link such a small mule to waterpower would be charged extra. To another customer they wrote "In respect to what number of Spindles may be most profitable is very Dificult to fix as what was thought best only two years ago is now thought too small. 216 is made to run as light now as 144 used to do then, though it cannot be said that it will do so much more work in proportion to the number of spindles. We are making now from 180 to 288 Spindles."[11] Bigger machines, bigger factories, bigger operations: by multiplying work in devices, industrialists were learning the lessons of mass production.

One popular version of the Industrial Revolution focuses on changing energy sources, as steam engines were applied to more and more machines in the eighteenth century, while textile production was mechanizing.[12] In 1765, after the Duke of Bridgewater opened a canal to run from his coal mines to Manchester, the price of coal in the town dropped by half.[13] So what role did steam power play in textile mechanization? The heroic inventor in this sector was James Watt, who was improving the eighteenth century's innovations in steam power with a separate condenser – and his business partner Matthew Boulton made sure everyone knew it. Watt had studied and worked as a mathematical instrument-maker, first in Glasgow and then London – a profession that owed much to the clockworks of which the Indian Mughal courts were so fond. Praise heaped upon Watt has declared, "After Watt, steam came of age and powered the industrial revolution." James Watt and his steam engines did have revolutionary effects. But it was a complicated story. For example, Watt "was cool to the high pressure engine." It was not until

[10] Mike Williams, with D. A. Farnie, *Cotton Mills in Greater Manchester* (Lancaster: Carnegie Publishing Ltd., 1992), 51–52; Margaret Jacob, *The First Knowledge Economy: Human Capital and the European Economy, 1750–1850* (Cambridge and New York: Cambridge University Press, 2014), 92, 95.

[11] McConnel & Kennedy to Saml. Pearson, Feb. 11, 1795; idem to Taylor & Haywood, Feb. 28, 1795; both, Letter Book 11 Feb. 1795–31 May 1796, pp. 1, 5.

[12] Richard L Hills, *Power in the Industrial Revolution* (Manchester: Manchester University Press, 1970); E. A. Wrigley, *Energy and the English Industrial Revolution* (Cambridge and New York: Cambridge University Press, 2010); Emma Griffin, *A Short History of the British Industrial Revolution* (New York: Palgrave Macmillan, 2010); Robert C. Allen, *The British Industrial Revolution in Global Perspective* (Cambridge and New York: Cambridge University Press, 2009).

[13] Maw, *Transport and the Industrial City*, 245.

after his patents expired in 1800 that high-pressure steam engines really began to work. By then, of course, textile industrialization was already well underway.[14]

Before Watt, steam engines had been used mostly around coal mines – they were powered by the coal and were used to pump water out of the mines. By burning coal, they heated water into steam that expanded up into a cylinder, pushing up a piston at one end of a giant beam, which pushed down the other end of that beam. When the steam cooled, it created a partial vacuum that pulled the piston back down the cylinder, lifting the other end of the beam, and with it the water from the mine. Watt patented a system of gears to transform this up-and-down motion into the rotational movement of a wheel, while waterfalls were still directly turning wheels in places like Quarry Bank. Another of his patents described his separate condenser, which removed the hot steam from the cylinder to create the vacuum. Separating the steam and keeping it hot made it easier to re-heat the steam for the next cycle. This cut the coal consumed by his steam engines to a third of the requirements of earlier engines.[15]

Watt's separate condenser was not too popular with Manchester's cotton spinners, however, who usually pursued cheaper engines. In fact, many cotton spinners installed steam engines only to raise water to the waterwheels that actually turned the machinery. Manchester's first pur-pose-built cotton mills rose in the 1780s, at least fifteen years after the opening of the Bridgewater Canal, so the link between coal prices and textile mechanization was not so tight that one followed the other as the price of coal plummeted. Peter Drinkwater, a one-time fustian merchant who built Piccadilly Mill in 1789, was famed as the first to apply steam power directly to preparatory machinery – but his 144-spindle mules were still "entirely hand-operated."[16] (see Figure 4.2.) Arkwright's own spin-ning mill in Manchester – built before 1782, on a grand scale to show that "the most powerful man in the cotton trade had built it" – was outfitted with a steam engine intended to run machinery, but since it consumed

[14] Milton Kerker, "Science and the Steam Engine," *Technology and Culture* 2, no. 4 (Autumn 1961), 387; D. J. Bryden, "James Watt, Merchant," in Denis Smith, ed. *Perceptions of Great Engineers: Fact and Fantasy* (London and Liverpool: Science Museum and the University of Liverpool, 1994), 9–10; Robert Fox, "Watt's Expansive Principle in the Work of Sadi Carnot and Nicolas Clément," *Notes and Records of the Royal Society of London* 24, no. 2 (Apr. 1970), 238.

[15] Ibid.; A. E. Musson and E. Robinson, "The Early Growth of Steam Power," *Economic History Review*, New Series, Vol. 11, no. 3 (July 1959).

[16] Williams and Farnie, *Cotton Mills in Greater Manchester*, 48–52, quotation at 52; Andreas Malm, *Fossil Capital: The Rise of Steam Power and the Roots of Global Warming* (London and New York: Verso, 2016), 55, 78, 89; Maw, *Transport and the Industrial City*, 211–13; Musson and Robinson, "Early Growth of Steam Power," 424–25.

five tons of coal each day, in the end it was used only when needed to raise water to a large waterwheel. The power-producing technologies of water-wheels and coal-burning steam engines existed side-by-side, and they drew on one another.[17]

Nonetheless chimneys were rising and men sought alternatives as the business expanded – some pirated the separate condenser idea, and Boulton and Watt sued infringers throughout the 1790s. The smoke-stacks and red-brick chimneys through which steam engines poured their waste smoke were among the marvels of Manchester. Other engineers studied Watt's designs, and made improvements, and stood ready to sell competitive wares when Watt's patent expired in 1800. Matthew Murray, for example, who had worked on flax-processing machinery with John Marshall in Leeds, built steam engines entirely out of iron while Boulton and Watt still made many components out of wood. He also sold con-densers separately from engines, and thus avoided litigation. In 1790, Boulton & Watt had sold only twelve engines to cotton mills: many mill owners, including Arkwright himself, found coal too expensive. The cost of operation was even higher for the cheaper Savery- or Newcomen-style engines that skipped the separate condenser. In 1800, the total number of steam engines that Boulton & Watt had sold to cotton mills numbered 84 – compared to around 1,000 factories still powered by water – and "all the major technical breakthroughs in cotton-spinning [were] originally developed for other forms of power." The majority of textile mills found waterpower cheaper, and stuck to it – into the 1830s.[18]

The Rochdale Canal Act of 1794 was one reason that Manchester's cotton manufacturers lingered behind the leading edge of steam engine development. The Canal Act allowed the factories built along its banks (and later, on canal branches, too) to draw the canal's water. They could extract water to condense in their steam engines and then return the water to the canal. This provision helped concentrate factory development to the canal networks of the city, a pattern that persisted into the mid-nineteenth century. It also influenced the choices its manufacturers made regarding power sources. It made cheap the operation of low-pressure steam engines, which used water for condensing the steam. In 1797, McConnell & Kennedy bought land for a mill along the path of the Rochdale Canal (see Figure 3.1). Earlier Manchester cotton firms rented room and turning (space in a factory building plus the power of its wheel), but McConnell & Kennedy built their mill for their own use. They also bought a low-pressure

[17] Christopher Aspin, *The Water-Spinners* (Helmshore: Helmshore Local History Society, 2003), 72–75, quotation at 72.

[18] G. N. von Tunzelmann, *Steam Power and British Industrialization to 1860* (Oxford: Clarendon Press, 1978), 183; Malm, *Fossil Capital*, 55–56, 78, 89.

Figure 3.1 Cotton mills of McConnell & Kennedy, on the banks of the Rochdale Canal, in the Ancoats neighborhood of Manchester, built from 1798 to 1912. Courtesy of Alberto Manuel Urosa Toledano/ Moment Open/Getty Images.

engine and stopped relying on the horses that had previously powered their spinning mules. Canal water on tap made less efficient engines cheaper to use than the more efficient high-pressure engines that would become available in the nineteenth century.[19]

The popular history therefore overstates the importance of steam engines to the industrialization process. It is true that steam engines supplied power

---

[19] Finding Aid, McConnel & Kennedy Papers; Williams and Farnie, *Cotton Mills*, 30–35, 48–52; Kenneth C. Jackson, "The Room and Power System in the Cotton Weaving Industry of North-east Lancashire and West Craven," *Textile History* 35, no. 1 (2004): 58–89; Peter Maw, Terry Wyke, and Alan Kidd, "Canals, Rivers, and the Industrial City: Manchester's Industrial Waterfront, 1790–1850," *Economic History Review* 65, no. 4 (Nov. 2012), 1503–10; Maw, *Transport and the Industrial City*, 211–13; Malm, *Fossil Capital*, 55–57.

to a wide range of processes. They were applied to general purposes while textile machinery was specific down to the type of fiber a machine would process. But even in the general economy, steam power did not triumph until the second quarter of the nineteenth century. As late as 1830, steam and waterpower provided roughly the same quantity of energy to the economy (about 165,000 horsepower each). Steam engines are therefore too late to explain the origins of the Industrial Revolution. Textile industrialization took place well before the steam engine was economically important. In addition, most steam engines themselves were bespoke, custom-designed for specific needs – they were themselves artisanal more than industrial artifacts. Men sent along to set them up usually traveled by horse-drawn coaches, as textile machinery went in boxes on wagons and boats. In the case of textiles in particular, steam engines were adopted only sporadically.[20]

We will therefore return to coal and steam engines in the nineteenth century, as they begin to matter more. For now, let us turn back to local history as a window to understand a different set of inputs to industrialization, including both raw materials and workers. The sources of fiber to spin into yarn in these factories matter as much to the story of industrialization as do the machines. The Industrial Revolution that turned a rooted, veteran textile industry into the world's primary case of mass production relied on the supply of cotton to Britain, which expanded in the last two decades of the eighteenth century. The connection between cotton plantations in the American South and the cotton mills of Manchester tightened even further in the nineteenth century, and the links passed through Liverpool. At the same time, the workers who operated the machines in Manchester also brought familiar forms, their experiences and customs, to the changing industrial condition, and made something new out of what they found: a class identity. The sources of cotton, and the making of the English working class, both contributed to the success of the industrial system. These contingent contributions enlarge our view beyond individual machines, and beyond even Manchester.

Greater Manchester

The immediate environs of Cottonopolis begin to widen the perspective. As Manchester and its output grew, hinterland cotton towns came to specialize in particular products. Bolton, where Samuel Crompton had

---

[20] von Tunzelmann, *Steam Power and British Industrialization*, 4–5, 183; Joel Mokyr, *The Enlightened Economy: An Economic History of Britain 1700–1850* (New Haven: Yale University Press, 2012), 96; Finding Aid, Archives of Soho [Boulton & Watt], Birmingham Central Library, Birmingham, UK.

invented his spinning mule, shifted from fustian production and became a cotton-spinning village. Its mills used long-staple cotton and spun fine yarn for the quality trade. It remained a center for a surrounding community of fine weavers as well. Oldham's population, on the other hand, grew rapidly during the period between 1775 and 1821. It became a stronghold of handloom weavers, weaving machine-spun yarns, before that trade disappeared into mechanized weaving a few decades later. At that point, its workers turned to spinning coarse yarns. In the nineteenth century, each town's yarns were well known in the trade: 60s were Bolton counts while 32s were Oldham counts (higher numbers meant finer yarns). Each town also had a characteristic form of business organization. Individual manufacturers owned Bolton's mills, and the town had its own satellite villages. In Oldham, joint-stock spinning companies rented space and turning within larger mill buildings. They concentrated on volume production of thicker yarn. By 1841, their workers earned the highest gross pay in the industry.[21] The differences between Bolton and Oldham show that while factories had gathered preparation and spinning processes under one roof, the divisions of labor characteristic of proto-industrialization were visible in new forms – at the level of villages rather than individual households.

Larger towns also specialized, and these clusters likewise represented the larger structural changes of industrialization. While Manchester became the center of cotton manufacturing, its mills separated from merchant activity and the overseas commerce associated with the port at Liverpool. Today, economists describe economic activity in terms of primary, secondary, and tertiary sectors – the first is agriculture or mining (extraction generally), while the second is manufacturing, and the third is service, including commerce. Before industrialization, theorists of mercantilism equated capitalism and commerce. Merchants even coordinated the putting-out of work and materials into people's homes. They invested in raw materials but not in what Karl Marx would later call the "means of production." Instead, in the classic case, workers owned the devices they used to turn raw materials into finished goods. The worker supplied what today would be called fixed capital (his home or workshop and the tools in it) and the merchant supplied the operating capital (the raw materials). In practice, it was more complicated than that. Labor costs are operating capital, and workers supplied it. Some putting-out merchants supplied looms and some traders owned shares in ships as well

---

[21] Williams and Farnie, *Cotton Mills in Greater Manchester*, 29–37; George Unwin, *Samuel Oldknow and the Arkwrights: The Industrial Revolution at Stockport and Marple* (New York: Augustus M. Kelley, 1968; orig. pub. Manchester University Press), 15–16.

as the working capital invested in raw materials and cargoes. But economists view merchants and manufacturers as working in different sectors,
with different types of investments. And the manufacturing sector was
itself both a product of and a contributor to industrialization.[22]

When Adam Smith (1723–1790) published *The Wealth of Nations* in
1776, he signaled the transition from mercantilism to this newer body of
economic theory undergirding free markets and capitalism. He celebrated
the benefits of dividing labor into tiny repetitive production tasks, exem-
plified in his book in the making of pins: the shaft and head made
separately, and sharpened and put together in further steps. In this way,
Smith described fabrication divided into parts, tasks ripe for replacement
by machines. In the same work, he criticized earlier theorists – of mercantilism, which he called "the mercantile system" – for not recognizing
that wealth could be grown through industry. He celebrated mass production and provided a rationale for distinguishing it from commerce.
Commerce still existed, of course. Merchants bring knowledge and networks of connections to the trade. A manufacturer should know what his
buyers want, but there are many buyers of shirts around the world, and
the right merchant will know what goods particular customers desire.
Merchants learn which makers have what goods on offer: fine dress shirts
are usually made by different firms than produce casual clothes or sports
gear. Matching makers and consumers is one way the service sector
makes its money. Distinguishing the work of commerce from manufacturing that invested in machines, sourced raw materials, hired workers,
and expected a return only over a long period of operation was one of the
conceptual transformations of industrialization.[23]

Commerce and Cotton

Commerce and manufacturing separated not only conceptually but also
literally in the 1780s, with the spread of cotton spinning on Arkwright
frames. Textile men from Manchester made their way to Liverpool and
specialized in buying cotton for sale to the spinning mills. They became
a new profession called "cotton brokers" and their business lodged at the
port, at Liverpool, while Manchester became associated with the grubbier
manufacturing side of the cloth business. The two were linked by canals
(and later, by railroads), but they were forming into two different sectors
of the one cotton industry. The distinction between them is now the

---

[22] John C. Black, *Dictionary of Economics*, 2nd ed. (Oxford and New York: Oxford
University Press, 2002), s.vv. "capital," "capital-labour ratio," "fixed cost," "sector."
[23] Adam Smith, *An Inquiry into the Nature and Causes of the Wealth of Nations* (London:
W. Strahan and T. Cadell, 1776), book i, chapter 1, and book iv, chapters 1–8.

sectoral distinction of economists, but then appeared more clearly in the relative social status of the two trades: In the nineteenth century, they would commonly be known as the Liverpool Lord and the Manchester Man. Overseas commerce was a little more gentlemanly. Manufacturing took effort, the kind of striving that belonged lower in the ranks. The appearance of economic sectors meant that merchants and manufacturers specialized, and so did the business of the towns.[24]

Liverpool had grown with the slave trade. The port city functioned as a node in Britain's Atlantic commerce, which stretched from the west coast of Africa to the American colonies. During the eighteenth century, Liverpool rose from fifth to third among the British ports – London, Newcastle, then Liverpool. Trading networks and local infrastructure accompanied this rise: Liverpool's Old Dock dates from about 1710, the Dry Dock opened after 1717, and the Salthouse Dock was commissioned in 1734 and opened in 1753. Construction began on George's Dock in 1767, and in 1788 the King's Dock opened, followed by the Queen's Dock in 1796. In 1802, the West India Docks were opened and the first stones for the London Docks were laid. Between 1816 and 1821 grew the Prince's Dock, just north of George's Dock. A century later, *Titanic* sailed from Liverpool. These infrastructural projects for ocean-going vessels were matched by the city's connections into the interior of Lancashire: the Sankey Canal in 1758 linked the Mersey to coal mines, just as Bridgewater's canal did a few years later, the latter of course passing through Manchester before reaching Liverpool and the sea. In 1774, the Leeds and Liverpool Canal opened, tying Yorkshire and its cloth markets to Liverpool and the Atlantic Ocean. In 1822, that link from Yorkshire to the Atlantic world added a connection to Manchester.[25]

Liverpool specialized in the "American trade," on the mercantilist model. Before industrialization, the American colonies had supplied raw materials – tobacco especially, but also rice, cod, and sugar – for the imperial metropolis to manufacture or reexport. The colonies had also served as a market for metropolitan goods. Britain's American settlers had been vital consumers in the East India Company's textile

---

[24] Stanley Chapman, "The Commercial Sector," in Mary B. Rose, ed., *The Lancashire Cotton Industry: A History Since 1700* (Preston: Lancashire County Books, 1996), 65; Thomas Ellison, *The Cotton Trade of Great Britain, Including a History of the Liverpool Cotton Market and of the Liverpool Cotton Brokers' Association* (London: Effingham Wilson, Royal Exchange, 1886), 167–68, 175; Isabella [Mrs. George Linnaeus] Banks, *The Manchester Man* (London: Hurst and Blackett, 1876).

[25] Chapman, "Commercial Sector," 65; Henry Smithers, *Liverpool, Its Commerce, Statistics, and Institutions with a History of the Cotton Trade* (Liverpool: Thomas Kaye, 1825), 169–74; Cookson, *Age of Machinery*, 50.

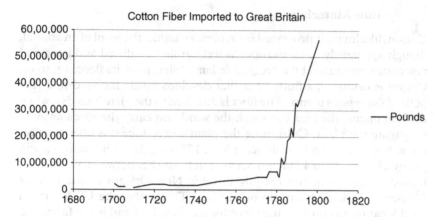

Figure 3.2 Cotton imported to Great Britain in the eighteenth century demonstrates the importance of raw material supplies to keep the spinning machines of the Industrial Revolution at work. Source: Baines, *History of the Cotton Manufacture in Great Britain* (1835).

business, especially crucial to British commerce when the imported cotton cloth was restricted or prohibited in England itself. The colonists had also bought millions of the people kidnapped and packed into ships on the west coast of Africa and sold to work on the American and Caribbean sugar plantations. Cotton fiber came late to this story: the importation of cotton to Liverpool in 1770 amounted to no more than 6,000 "small bags," according to one authority, and none of these came from North America. The quantity soon jumped. Britain imported nearly five million pounds of the fiber each year between 1771 and 1775 and an average of 6.7 million pounds in each of the next five years. As Arkwright-style mills spread across the countryside, Britain nearly doubled its annual consumption, each year, for several years after 1781 (see Figure 3.2). With these increasing quantities of supplies and the rising demand of spinning machines and their mills, and most cotton for the hungry spindles arriving in Liverpool, the Lancashire port city outpaced London, in the American trade at least. By 1795, Liverpool had become the main cotton importer of Britain.[26]

---

[26] S. G. Checkland, "American versus West Indian Traders in Liverpool, 1793–1815," *Journal of Economic History* 18, no. 2 (1958), 141; Ellison, *Cotton Trade*, 166, 170–71; Joseph E. Inikori, *Africans and the Industrial Revolution in England: A Study in International Trade and Economic Development* (Cambridge and New York: Cambridge University Press, 2002), 192–93; Edward Baines, Jr., *History of the Cotton Manufacture in Great Britain* (London: H. Fisher, R. Fisher, and P. Jackson, 1835), 215.

Raw Materials

Cotton, like linen, is processed from plants – unlike the wool of an animal, though apparently some Europeans were at first confused and imagined raw cotton produced by a "vegetable lamb" that grew its fleece on trees. Cotton is cellulose, a fluffy fiber that develops inside the seed-pod (the boll) of *Gossypium* plants. The fiber is attached to the plant's seeds – when the boll opens, the fuzz can catch the wind and carry the seeds away to propagate the plant. Cultivating the plant to produce this fiber requires thick, heavy soils, and sunshine, at least 175 long days without frost, and plenty of rain – 24 to 47 inches per year. It flourishes between 30° and 37° latitude – lines that run roughly through New Orleans in the Gulf of Mexico and the northern tip of Virginia. This makes the US South an ideal location to cultivate the plant for its fiber. England is too far north. Its growing season is too short. So British spinners had to use cotton from elsewhere. This suited mercantilist ideology and practice, as cheap raw materials from the global South fed manufacturers' mills at home. However, the sources of the fiber shifted during industrialization, as the quantities imported exploded.[27]

Before the Industrial Revolution, cotton had been brought to the British Isles from diverse regions. India had its own supplies, cultivated for use in its own industry. By 1200, however, at least small quantities of cotton had made their way to England through Venice. As we know from Chapter 1, the English only slowly learned to use the new fiber in bedding and clothing. In the early 1500s, cotton imported through the Levantine trade was used for candlewicks and the stuffing of quilts. By 1611, some fustians were manufactured in England and by the end of that century the cotton-linen mixture was a major domestic product, manufactured mostly in Lancashire, for sale in France, Spain, Holland, Germany, and Italy. Cotton's price rose with the demand. Between the 1720s and 1780, cotton grown in Syria and Cyprus tripled its price in Amsterdam. But Manchester's textile business received most of its raw cotton from the West Indies, colonial plantations in the Caribbean Sea, until the 1780s. At that point, and just a few years after achieving their political independence, the new United States suddenly began to grow cotton.[28]

The myth tells us that customs officials in Liverpool were so shocked by the arrival in 1784 of eight bags of cotton from the new United States that

---

[27] Sven Beckert, *Empire of Cotton: A Global History* (New York: Alfred A. Knopf, 2014), 22–23; Bruce E. Baker and Barbara Hahn, "Cotton," *Essential Civil War Curriculum* (2015).

[28] Giorgio Riello, *Cotton: The Fabric That Made the Modern World* (Cambridge and New York: Cambridge University Press, 2013), 15, 74–75, 253–56; Wadsworth and Mann, *Cotton Trade*, 20–21.

they seized them for quarantine. If the bags had originated on the loyal plantations of the West Indies, it was illegal for the independent Americans to intervene that way in the coastwise trade.[29] But the transatlantic commerce that after this moment took off, expanding across the nineteenth century, was no fairy tale. Industrialization in Britain and cotton slavery in the United States grew up together; factory and plantation masters were linked by abolitionist contemporaries as "lords of the lash and lords of the loom." American cotton production relied on the plantation form of production as the United States expanded in the nineteenth century. In the Mississippi River Valley, steam-powered boats linked ever-more-productive slave plantations to world markets through the Gulf of Mexico. By 1860, cotton totaled 60 percent of the value of exports from the United States, and the fiber went mainly to Britain. This was new cargo: the British North American colonies had produced different commodities. The plantation system of production developed first in sugar cultivation was applied to cotton production only after American independence.[30]

The importance of cotton to the Industrial Revolution has recently become the subject of heated disagreement among historians. Part of the definition of industrialization concerns "the getting and working of raw materials," and the regularization of both their flow and their traits.[31] As we know, other fibers mechanized parts of spinning, but the scale of output in woolens and worsteds, linens and silks, was simply not as revolutionary as in cotton. And the hundred-fold increase in raw cotton imported between 1772 and 1841 was both cause and effect of the rapidly increasing output, of the changing machinery between raw and finished conditions.[32] To pose the question more precisely, was cotton grown on the slave plantations of the American South crucial to the Industrial Revolution?[33] Not quite. Those mythical eight bags that landed at Liverpool in 1784 came too late to have caused the jenny, water-frame, or the mule that revolutionized spinning. However, expanding the new systems relied on having raw materials to feed more spindles, and the fiber

---

[29] Ellison, *Cotton Trade*, 81; Smithers, *Liverpool*, 124.

[30] Robert F. Dalzell, *Enterprising Elite: The Boston Associates and the World They Made* (Cambridge, MA, and London: Harvard University Press, 1987), 202; Inikori, *Africans and the Industrial Revolution*, 192–93; Stuart Bruchey, *Cotton and the Growth of the American Economy, 1790–1860: Sources and Readings* (New York and Chicago: Harcourt, Brace & World, Inc., 1967), tables 3A and 3J.

[31] David S. Landes, *Unbound Prometheus: Technological Change and Industrial Development in Western Europe from 1750 to the Present* (Cambridge: Cambridge University Press, 1969), 1.

[32] Inikori, *Africans and the Industrial Revolution*, 78–79.

[33] Alan L. Olmstead and Paul W. Rhode, "Cotton, Slavery, and The New History of Capitalism," *Explorations in Economic History* 67, no. 1 (2018): 1–17.

fit the American context, with its ready-made plantations and legal structures supporting heritable conditions of slavery, to provide labor bound to
the work.

Economic historian Joseph Inikori has framed the question in terms of
the importance of Africans to the British Industrial Revolution. By tracing
the growth of Britain's international trade between 1650 and 1850, he
calculated how much of that growth was a function of Atlantic commerce,
which depended on the forced labor of Africans. His version of the
Industrial Revolution incorporates regional English history, fabrics
other than cotton, shipping and finance, and he also provides a series of
international comparisons designed to examine the causes of industrialization in northern England's textile production. His precise measure-
ments of the contributions of Africans, both continental and diasporic,
demonstrate conclusively their roles, as producers and consumers, to
Britain's industrialization.[34]

For new machinery to result in mass production of goods for distant
consumers, then, mass production of the raw materials was also needed.
It is easy to imagine alternative technologies, and important to remember the intricacy of networks that produced the one at the center of our
story – the mechanization of textile production in the last decades of the
eighteenth century and its increase in the nineteenth. At many points,
a different input, or device, or outcome, was possible. Still, the task of
this book is to explain what did happen. The Industrial Revolution did
depend upon commodity production, and the agricultural work of
enslaved people, for the mechanization of textile production in
Manchester and its hinterlands. Enslaved agricultural workers produced cotton fiber from plants in the Americas, which the masters
who had bought their bodies and the merchants who financed their
work shipped to England in order to make textiles that imitated Indian
cloth. British textiles made from that cotton went to Africa to finance the
purchase of more muscle, and the cloth made in Britain also sold in
America. The world system of commerce that underlay the Industrial
Revolution in Manchester was global in scale and its networks wove
across the Atlantic.[35]

Industrial capitalism that relied upon slave-grown raw materials provides yet another example in which new systems drew on older forms even
as they transformed or destroyed them. The one-time colonial relationship that had developed the commerce between Britain and the United
States shifted its basis to cotton after independence. The plantation
production system itself had developed in mercantilist times. It preserved

---

[34] Inikori, *Africans and the Industrial Revolution*, xvii–xix.   [35] Ibid.; Riello, *Cotton*, 86.

the household as a farm, a place that originated goods, even as it began growing raw materials for industry. On the plantations that cultivated raw materials for the mills, workers lived on site, and masters or managers often did too. The cotton plantation was a modern agro-industrial enterprise couched in an older form of household production. Planters viewed themselves in almost medieval terms, as paternalists responsible for both the output and conduct of their inferiors, whether slaves or wives or children. Unfree labor was itself a variant of the patriarchy of household and master-apprentice production. Slaves, indentured servants, and apprentices: a spectrum of unfreedom spanned industrialization. The plantation sources of industrialization, along with merchant capitalism and the reorganization of world trade effected through the East India Company, show how industrial capitalism drew on economic structures that already existed.[36]

India and Imperial Ideas

As Giorgio Riello argues in his history of *Cotton*, the Industrial Revolution was the epiphenomenal effect of changing patterns of world trade, rather than the result of individual genius or institutions, or the local prices of coal and wages. The Industrial Revolution was indeed a response to Indian imports, but the relationship between England and India was also changing as industrialization commenced. India's Mughal princes had never understood why a trading company needed an army, but the East India Company showed its reasons when it used guns to secure control of Bengal in the 1757 Battle of Plassey, then tightened its grip in 1764. The EIC then took over the diwani, an indigenous system of tax and civil administration, and its proceeds were vast. In 1770, when famine killed a third of Bengal's population, still the EIC collected 15 million rupees. The Company forced down the price it paid to weavers and seized their goods for sale. Bengali weavers revolted – some were reported to have cut off their own thumbs rather than make goods for the Company. Incorporating native trading networks into the EIC was tricky and took time: into the nineteenth century the EIC was still struggling to control local markets and fairs. Not until 1858 did the Crown take over from the Company, but Riello has described the centuries preceding the 1858 Raj as an English "apprenticeship" in the business of

---

[36] Sidney W. Mintz, *Sweetness and Power: The Place of Sugar in Modern History* (New York: Penguin Books, 1986; orig. pub. 1985); Jared Ross Hardesty, *Unfreedom: Slavery and Dependence in Eighteenth-Century Boston* (New York: New York University Press; 2016).

empire.[37] Britain was twisting together the threads that afforded its global rule.

While Parliament in the eighteenth century had given British textiles a decisive advantage over Indian imports, the cottons of the subcontinent still competed elsewhere around the world. As early as 1782, British merchants were sending samples of the cloth being made in Manchester to India and Bengal as a warning about new competition. In response, the East India Company began sending better Bengali goods to market. Since the Calico Acts, the EIC had sold Indian cottons in Africa and North America, so the contest between British and Indian textiles took place there first, in the Atlantic trade. The independent American states were already consuming much more British-made cotton cloth than Indian imports: between 1784 and 1786, the proportion was 88 percent British and 12 percent Indian cottons; ten years later, Americans received only 4 percent of their cottons from India and the percentage declined thereafter. In Africa, however, the EIC was still selling Indian textiles in Africa to buy slaves. As the eighteenth century ended, the economic system of Atlantic trade shifted to cotton, and raw materials from the plantations of the southern United States were made into cotton cloth in Lancashire, to compete with the Indian wares around the world. Meanwhile, a respite in England's wars with revolutionary France allowed Manchester goods to be sold in Europe after 1802, where they competed with Indian cloth and expanded the British cotton industry. In 1799, there were 51 cotton-spinning firms in Manchester and by 1811 the number had more than doubled to 111.[38]

Riello is right, then, that industrialization was a result, more than a cause – at least at first. It was not the institutions nor the character of great British men that inspired mechanization of spinning, although the existing textile business and the protective laws did provide people and firms and know-how, institutions and trade networks, and of course

[37] Giorgio Riello, "The Indian Apprenticeship: The Trade of Indian Textiles and the Making of European Cottons," in Giorgio Riello and Tirthankar Roy, eds., *How India Clothed the World: The World of South Asian Textiles, 1500–1850* (Leiden and Boston: Brill, 2009); Riello, *Cotton*, 6–7, 211; Sudipta Sen, *Empire of Free Trade: The East India Company and the Making of the Colonial Marketplace* (Philadelphia: University of Pennsylvania Press, 1998), 74, 89, 90, 114–15, 120, 134, 163; Indrajit Ray, *Bengal Industries and the British Industrial Revolution (1757–1857)* (London and New York: Routledge, 2011), 9–10.

[38] Inikori, *Africans and the Industrial Revolution*, 428, 447, table 9.9; H. V. Bowen, *The Business of Empire: The East India Company and Imperial Britain, 1756–1833* (Cambridge: Cambridge University Press, 2006), 242–45; Prasannan Parthasarathi, *Why Europe Grew Rich and Asia Did Not: Global Economic Divergence, 1600–1850* (Cambridge and New York: Cambridge University Press, 2011), 131; Williams and Farnie, *Cotton Mills in Greater Manchester*, 19.

privileged positions in home markets. But technological change reverb-
erated. As England industrialized, under protective cover, its products
began to compete with the Indian fabrics that clothed the world. The
industry that materialized in changing technological systems had effects
on world trade patterns as well as emerging from them. Thus the
Industrial Revolution provides a satisfying example of the relationship
between technological and contextual change. As new machines arose in
response to the demands of world markets, so did the products pouring
from those devices change the commercial relationships.

Divisions of Labor

The networks of Cottonopolis extended across the globe, from the Indian
Ocean to the North American frontier. Shift the lens, though, and industrialization also had effects on individuals and households, not only on the
plantations of the American South and the villagers of Bengal but also for
the cotton mill workers of Lancashire – their livelihoods, of course, but
also the organization of their families and households. As industrialization
stabilized and expanded, more workers entered the mills. Pauper children
remained a labor resource, but growing mills and mass production drew
people into jobs, and the concentration of the cotton industry pulled
many of them to Manchester. The poor children consigned to the parish
had worked to power the mills as the system was developing along waterways in the countryside. As the factories coalesced in the towns and
villages of Lancashire, a new system assembled.[39]

There were multiple sources of factory labor, as there were many
historical trajectories that contributed to the development of the factory
idea. Apprenticeship arrangements lasted well into the nineteenth century. Parliament tried to regulate child labor from 1802 through the
Factory Act of 1833, with limited success. Capitalists kept using the old
system, claiming that they needed child labor to turn profits. While the
recent quantitative study of their accounting methods has found little
evidence for this point, they used economic arguments that were widely
accepted in the nineteenth century.[40] Factories like Quarry Bank, with its
apprenticed workforce, demonstrate the persistence of older social structures within the new mechanized and powered production systems.

---

[39] Katrina Honeyman, *Child Workers in England, 1780–1820: Parish Apprentices and the
Making of the Early Industrial Labour Force* (Aldershot, England and Burlington, VT:
Ashgate Publishing, 2007).

[40] J. S. Toms and Alice Katherine Shepherd, "Creative Accounting in the British Industrial
Revolution: Cotton Manufacturers and the 'Ten Hours' Movement," 5–8, 21–22,
26–33.

A master weaver who produced cloth at home and trained apprentices in a domestic workshop could be classified as an industrial capitalist due to his fixed sunk investment in looms, even though his workers lived in the shop. Individuals working at home, in large and highly capitalized workshops, in water-powered factories like Samuel Greg's and others that used steam engines, all coexisted. Industrial and domestic and artisanal modes of production were not entirely distinct from one another.[41]

Arrangements for mill workers had taken several possible forms from the start. Arkwright himself employed the wives and children of local lead miners in his Cromford mill, and these worked on contract, for specific periods of time and specific pay. As the eighteenth century ended, men like Samuel Greg and Samuel Oldknow also shifted to contract workers, though neither thoroughly nor rapidly. Greg had close connections at the Liverpool poorhouse that kept him supplied with workers, and not until 1847 did Quarry Bank shift to a voluntary workforce, though free workers had mingled with apprentices since at least 1783. Contract workers could work in the same mill as pauper apprentices but their bosses had fewer responsibilities to them. Cutting labor costs was easier from a non-apprenticed workforce, but getting workers to remote or rural locations could be tricky – that is why Greg built up the village at Styal after 1815. An urban setting had more jobs, more housing, and arguably more independence for laborers. Memoirs left by workers who lived through the Industrial Revolution illustrate their pleasure in the freedom of the new era, in wages and spending power, and in the consumer goods wrought by the new industrial world. Despite exploitation and their long struggle for political rights that we shall find in future chapters, people reveled in the possibilities.[42]

Free labor superseded the apprentice system bit by bit. Manchester's population increased by an order of magnitude between 1760 and 1830, growing from 17,000 to 180,000. The population tripled in just the fifty years leading up to 1794, as people seeking work streamed into the city "from the expropriated farms, the enclosed commons, the bankrupt shops, the ruined small 'manufactories'" across Britain. The work they found was mostly in the mills and mines. The housing they could afford was "offensive, dark, damp, and incommodious." Disease ran rampant in

---

[41] Stanley Chapman, *The Early Factory Masters: The Transition to the Factory System in the Midlands Textile Industry* (Newton Abbot: David & Charles, 1967), 37–38; Giorgio Riello, "Strategies and Boundaries: Subcontracting and the London Trades in the Long Eighteenth Century," *Enterprise and Society* 9, no. 2 (June 2008), 249.

[42] Rose, *Gregs of Quarry Bank*, 28–32, 78; Barbara M. Tucker, *Samuel Slater and the Origins of the American Textile Industry, 1790–1860* (Ithaca and London: Cornell University Press, 1984), 41–42; Emma Griffin, *Liberty's Dawn: A People's History of the Industrial Revolution* (New Haven: Yale University Press, 2013).

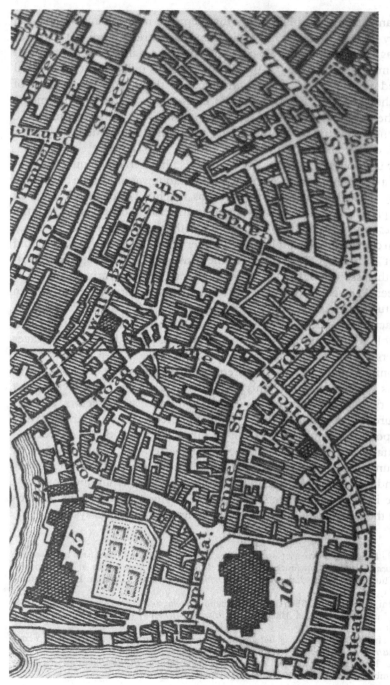

Figure 3.3 Map of Manchester and Environs (Swire, 1824), detail, shows the dense housing, built around nesting courtyards, that accumulated waste and impeded the free flow of air. Photograph by the author.

the dank cellars and blind alleys that kept the new urban dwellers and factory workers from access to light and air. Streets were not cleaned, drainage and lighting were barely known, and a dark muck ran through the dim streets. The smoke and dirt of the city's mills, added to the coal burned for heat at home, contributed to the overall impression of Manchester's filth. Black soot from coal smoke coated the tenements and the grand and dramatic red-brick mills alike.[43] Life was miserable in Manchester, but for many people, poor children included, their prior conditions had been even worse.

For their bosses, however, Manchester's filthy, cramped quarters provided a deep source of desperate workers. Labor was one of the benefits of closely packed urban industry, even as machines replaced workers in more phases of production. As textile industrialization consolidated, the business encompassed contradictions like these, characteristic of technological change. For example, early adopters of the Arkwright system could expect returns of 50 percent if they invested at the right moment. Nonetheless, four out of five of Manchester's manufacturing firms failed between 1780 and 1815, partly because their creditors had to respond to downturns with contraction, making business cycles worse.[44] These failures actually made starting up business cheaper: used machinery was available from bankrupt firms, mills rented room and power, and eager workers lived nearby. Factory masters with mills to keep running could subcontract some of the work (preparation and finishing, for example, as well as weaving) to merchants or households. The rapid growth of the industry also contributed to its articulation into distinct sectors, as manufacturers had begun to leave the commerce in raw materials to the Liverpool merchants. When villages like Oldham and Bolton became manufacturing centers satellite to Manchester, the growing city that surrounded the Ancoats industrial neighborhood shifted toward marketing and distribution for the manufacturers of the surrounding county.[45] These impulses and effects of changing technological systems demonstrate the complicated working out of a new technological system.

---

[43] Frida Knight, *The Strange Case of Thomas Walker: Ten Years in the Life of a Manchester Radical* (London: Lawrence & Wishart, 1957), 19–22, quotations at 21; Malcolm I. Thomis, *The Town Labourer and the Industrial Revolution* (London: B. T. Batsford, 1974), 48.

[44] Malm, *Fossil Capital*, 58–61; Stanley Chapman, *The Rise of Merchant Banking*, 10–11, cited in Boyd Hilton, *A Mad, Bad, and Dangerous People?* (Oxford: Clarendon Press, 2006), 23.

[45] Pat Hudson, *The Genesis of Industrial Capital: A Study of the West Riding Wool Textile Industry, c. 1750–1850* (Cambridge: Cambridge University Press, 1986), 14; Jane Humphries, *Childhood and Child Labour in the British Industrial Revolution* (Cambridge and New York: Cambridge University Press, 2010).

Textiles other than cotton present even more types of machine and labor arrangements. Silk throwing had mechanized in the eighteenth century, as Lombe's mill showed in Derby in 1719, and the model had spread and scattered across the country during the eighteenth century. The largest eighteenth-century silk mill was in Stockport and it had six engines and 2,000 workers. Meanwhile, putting out work into homes organized the production of worsteds in Yorkshire, while woolens retained a more artisanal, mixed agriculture-and-textile system. Fustian makers also put out work to domestic laborers, and that industry provided an important source of skills and capital to the industrializing, mechanizing production of cotton yarns. History and contingency account for this wide assortment of possibilities. Even after the factory production of cotton yarns had solidified into a package that could be sold and established in new locations, even after Arkwright, other methods survived, and thrived, in competition and combination with the factories that were still drawing labor from a range of possible sources.[46]

To add to the complexity of the picture, remember that all the above descriptions cover only spinning. Weaving remained a domestic occupation for fifty years after the mechanization of cotton spinning in Lancashire. A power loom had been patented in 1785, but it did not take hold in England for another three decades at least. John Kay's 1733 flying shuttle had multiplied the work of an individual weaver who tugged the piece of string that sent it flying, but it had not made the man obsolete. Quite the contrary: most firms that spun cotton yarn also employed a few dozens or hundreds of men at their looms. More than a few firms each paid a thousand men or more, for weaving cloth pieces in their homes, in the country and the villages surrounding the spinning towns. McConnell & Kennedy, for example, employed more than a thousand handloom weavers in 1816. In 1829, cotton-weaving handlooms outnumbered powerlooms by more than four to one: there were 55,500 power looms in England at the most, but almost 240,000 handloom weavers.[47] Factory spinning intensified and expanded the older system in which putting-out merchants coordinated the work of domestic producers, paid by the piece for the work they did. Between 1795 and 1833, the number of cotton

---

[46] Maxine Berg, *The Age of Manufactures: Industry, Innovation, and Work in Britain, 1700–1820* (Oxford: Basil Blackwell, 1985), 211–25.

[47] Malm, *Fossil Capital*, 69; Duncan Bythell, *The Handloom Weavers: A Study in the English Cotton Industry During the Industrial Revolution* (Cambridge University Press, 1969), 29–30; Cookson, *Age of Machinery*, 19.

handloom weavers who worked in their homes or workshops tripled in Britain.[48]

As eighteenth-century industrialization expanded into the nineteenth century, mechanized cloth production remained incomplete – whole sectors of production were still done by hand, at home, by men whose inputs now came from factories rather than their families. Samuel Oldknow's records once again reveal this system in action. Separate from his spinning operations, by 1786, he employed more than 450 weavers around Stockport. He specialized in fine muslins that were woven on a draw loom (a relatively new and expensive device that drew down the heddles of warp threads as well as raising them up), so he provided these machines, as well as their reeds and gears, to his weavers, for a fee. They carried their finished cloth to Oldknow's warehouse where they received the yarn from his mill for their next period of work. He provided weavers with both mule twist and water twist, yarns made on either Crompton-style spinning mules or Arkwright-style water-frames – another example of the technological systems that coexisted throughout industrialization.[49]

While machine spinning regularized yarn into counts, woven and finished cloth required more attention to fashion. Oldknow occasionally went down to London and visited the East India Company sales rooms. There he could see for himself what among the imports sold (probably to merchants for reexport) and what languished. The imported offerings of Mulmul handkerchiefs, of tanjeebs, addaties, cossaes, and nainsooks informed his own choices. He told his home-based weaving workforce what to make and the products included both fashionable fabrics and consumer goods. His workers made in their homes not only plain muslin to be sold in pieces, but also cord check and cord stripe, Bengal stripe, and ticking. Oldknow's system also produced specialty items, some of them quite finished articles: d'oyleys, napkins, towels, shawl collars, and cravats. He provided warps to the weavers, and often consolidated the making of warps, by hand, in small mills, as the correct warp was needed to weave a pattern, a check or stripe, correctly. He had some men weaving red-bordered handkerchiefs, for example, and a solid red border required red yarn in the warp as well as the weft. Customers and their

---

[48] Peter Kriedte, "Decline of Proto-Industrialization, Pauperism, and the Sharpening of the Contrast between City and Countryside," in Peter Kriedte, Hans Medick, and Jürgen Schlumbohm, *Industrialization before Industrialization: Rural Industry in the Genesis of Capitalism* (Cambridge and London: Cambridge University Press and Paris, Editions de la Maison des Sciences de l'Homme, 1981), 156.

[49] Unwin, *Samuel Oldknow*, 45–46; "Weaving Prices Book, 7 March 1791," Samuel Oldknow Papers.

discriminating tastes drove his business choices. Even mechanized production had to be attentive to consumer desires, and flexible to reflect shifting preferences.[50]

Across the industry, just as in Oldknow's accounts, domestic production and factory production coexisted and served one another. So did other technological systems. Workshops, homes, waterpower and steam engines, jennies and water-frames and spinning mules: the array of products, production systems, devices, and processes reveals the complex, contradictory nature of technological change. These incongruities included the workers in every part of the system. The guild-based apprenticeship structure provided unskilled operatives to the first mills, but were also the means of training those machine makers who were developing the new devices and production systems. The cotton industry used raw materials grown by enslaved laborers on one-time colonial plantations, which shows the persistence and expansion of household production at the same time that factory systems were solidifying. Such contradictions and the coexistence of so many technological constellations could mean conflict when workers felt their livelihoods, their control of workflow, or the organization of their domestic arrangements threatened.

Resistance

Worker dissatisfaction was, of course, older than textile mechanization, but its goals changed in the prism of the Industrial Revolution. The mythological Industrial Revolution is full of stories of resistance to the new technology, in which domestic producers went to mills to break machines that threatened their livelihoods. There were jenny-breaking riots in Blackburn in 1778 that drove James Hargreaves to move his operations to Nottingham where he started Hockley Mill, where he later adopted Arkwright's machines. There, Nottingham's hosiers also rampaged to oppose him. Home-based spinners broke machines all around the Lancashire hinterlands of Manchester in 1779. Machine-breaking, a protest against changing technology, remained one form of protest available in that place and time. In other cases, workers wanted better wages within the existing system – McConnell & Kennedy experienced this early in 1795, when their

---

[50] Finding Aid; "Callicoes and Other Goods for Sale at the East India House, March Sale, 1789;" "Warping Book," 12 Sep. 1787–28 Oct. 1791; "Warping Book," 12 Nov. 1787–23 Sep. 1793, all, Samuel Oldknow Papers; Unwin, *Samuel Oldknow*, 45–46.

"Journey-men Spinners made almost a Generall Turn Out for an Extravagent advance of wages." The strike was only resolved a month later when the firm gave in to their spinners' demands. In this case, strikers implicitly accepted the job and its technological systems of production – the struggle was over the terms. Nonetheless, the firm referred to its workers in guild terms: it was journeymen who turned out, not operatives or machine minders.[51]

   Both factory and domestic worker revolts relied on older traditions and prerogatives based on artisanal relationships to confront the perceived threats posed by new machinery and the new systems. This is another way that the industrial technological system drew on what already existed and yet also transformed or destroyed it, resulting in new structures. At the same time, however, popular uprisings against machines, mills, and factories became part of the industrial system – angry workers could spark mechanization as well as resist it – a process we shall see in the next chapter, in the further mechanization of mule spinning. In the next century, too, labor's interests would be codified and embodied in workers' organizations and trade unions, with the strike or work stoppage becoming the tool of resistance or negotiation. In the first decades of the new system and the last of the eighteenth century, resistance was as piecemeal and hybrid as the industry which it might protest.[52]

Politics and Religion

Manchester had a radical history, along with much of Britain's borderlands, the North of England and up into Scotland. Now, in the Age of Revolutions that sprang from American independence, political authority was in flux. Ideas of liberty had circulated through Europe in the Enlightenment writings of Rousseau, Voltaire, Hume, and Locke. Then came the French Revolution, which did more than disrupt English textile markets – it brought to brutal reality the crisis in inherited authority that accompanied the European Enlightenment. At first, the British could view the French as imitating the constitutional monarchy that ruled England after its own Glorious Revolution. A portion of the praise for the French example became full-blown republicanism that supported claims to universal political rights, as articulated in the work of Thomas Paine (1737–1809). His 1776 volume *Common Sense* had inspired British

---

[51] McConnel & Kennedy to G. Hannay, Nov. 12, 1795, Letter Book 11 Feb. 1795–31 May 1796, pp. 139–40; Chapman, *Early Factory Masters*, 46–48; Aspin, *Water-Spinners*, 55.

[52] Thomis, *Town Labourer*; John Stevenson, *Popular Disturbances in England, 1700–1832* (New York: Routledge, 2013; orig. pub. 1979).

North Americans to revolt for independence from Britain, and his 1791 *Rights of Man* brought these issues of representative government home to British soil. Nonetheless, the French destruction of its monarchy, and its replacement with a Republic, threatened the powerful interests that controlled English government.[53]

War between France and Britain began in 1793, and resulted in a crackdown on the activities of the Britons who supported the French Revolution too long or too loudly (these were generally called Jacobins). But the revolutionary impulse had arrived in Britain and homegrown radical groups sprouted in every level of society. Working people used these organizations to agitate for political rights. The London Corresponding Society wanted universal suffrage and annual Parliaments; its 3,000 members were artisans and merchants, and its founder had been a shoemaker. Organizations wishing to make government more responsive to people, using the idea of human rights, had become common in the Age of Revolutions. These institutions were found not only in London but also in the nodes and networks that spread across the country, including the North and including Manchester. Their members assembled to voice their grievances, and their gatherings created a group identity with both political and economic goals. They represented to Parliament a threat of Jacobin unrest, if not revolution. As a response to the workmen organizing to press for their rights, Parliament in 1799 and 1800 passed Combination Acts that outlawed trade unions and collective bargaining. Affirming the rights of the masters, of capital over labor, Parliament attempted to quash all collective resistance to the emerging industrial order.[54]

The Combination Acts outlawed the trade unions that had borrowed and transformed the old guild protections of patriarchal power. Guild purposes and privileges survived in other institutions, however, including those of masters turned industrial capitalists. Manchester's manufacturers formed new clubs, and at the same time were woven into social networks and existing community relationships in ways that directed the path of the industry. Manchester's General Chamber of Manufacturers, for example, founded in 1785, tangled with Parliament over a tax on

---

[53] Albert Goodwin, *The Friends of Liberty: The English Democratic Movement in the Age of the French Revolution* (London and New York: Routledge, 2016; orig. pub. 1979), 42; Ruth Mather, "The Impact of the French Revolution in Britain," British Library, 14 May 2014, www.bl.uk/romantics-and-victorians/articles/the-impact-of-the-french-revolution-in-britain (accessed 16 June 2017).

[54] 39 Geo. III, *c.* 81; 39 & 40 Geo. III, *c.* 106. Thompson, *Making of the English Working Class*, 17; Malcolm Chase, *Chartism: A New History* (Manchester and New York: Manchester University Press, 2007), 2.

fustian and received nationwide support in the form of petitions to the House of Commons supporting its efforts to protect domestic industry. Protesting against the fustian tax led Manchester men to unite into an organization, founded in the opposition to a tax, ready to represent the economic interests of their industry. No longer granted power through the guilds, the emerging industrial class assembled into new institutions, just as laborers were trying to do.[55]

Some of these political efforts came from an older radical tradition stretching back to the Civil War, a religious radicalism that drew some of its strands from Dissent from the established Church of England. Many major manufacturers, like the Gregs of Quarry Bank, were Dissenters from the Church of England. Religion played into politics: many Dissenters were also political Whigs, and the Conservative party ("Tories," many of whom were High Church Anglicans and therefore deeply devoted to the Church of England) feared for its own hold on local government.[56] While some of Manchester's new industrialists were deeply conservative, their party affiliation was less sure, as we shall see in the nineteenth century. Among the manufacturing class too were at least a few Jacobins, who opposed war with France and America, despite the revolutions that had taken place in each country, threatening the complacency of political elites. The traditional class structure was under stress. The French Revolution and the ideals from which it sprang – ideals of equality and representative government that came from Enlightenment – were challenging the customary social order. And some of these challenges came from the industrialization process itself.

Subordinate classes, freed from the paternalism of feudal tenantry, were giving way to a working-class identity antagonistic to the capitalist industrialists. Accelerating these distinctions and articulating the borders between them were two forces. The steam engine stood ready to change relations between capital and labor in the textile business. Its operation would express the fundamental conflict between labor and capital that became the basis of Karl Marx's theories, which also had effects. The observations and theoretical constructs of Karl Marx and Friedrich Engels, authors who categorized and analyzed the struggles occurring at the site of industrial production, would also sharpen the story.

[55] Knight, *Strange Case of Thomas Walker*, 29, 31; Cookson, *Age of Machinery*, 185; Witt Bowden, "The Influence of the Manufacturers on Some of the Early Policies of William Pitt," *American Historical Review* 29, no. 4 (July 1924).
[56] Knight, *Strange Case of Thomas Walker*, 23.

Suggested Readings

Beckert, Sven. *Empire of Cotton: A Global History*. New York: Alfred A. Knopf, 2014.

Bowen, H. V. *The Business of Empire: The East India Company and Imperial Britain, 1756–1833*. Cambridge: Cambridge University Press, 2006.

Foner, Laura. *Slavery in the New World: A Reader in Comparative History*. New York: Prentice-Hall, 1969.

Honeyman, Katrina. *Child Workers in England, 1780–1820: Parish Apprentices and the Making of the Early Industrial Labour Force*. Aldershot, England and Burlington, VT: Ashgate Publishing, 2007.

Humphries, Jane. *Childhood and Child Labour in the British Industrial Revolution*. Cambridge and New York: Cambridge University Press, 2010.

Inikori, Joseph E. *Africans and the Industrial Revolution in England: A Study in International Trade and Economic Development*. Cambridge and New York: Cambridge University Press, 2002.

Kidd, Alan, and Terry Wyke, eds. *Manchester: Making the Modern City*. Liverpool: Liverpool University Press, 2016.

Maw, Peter. *Transport and the Industrial City: Manchester and the Canal Age, 1750–1850*. Manchester and New York: Manchester University Press, 2013.

Williams, Mike, and with D. A. Farnie. *Cotton Mills in Greater Manchester*. Lancaster: Carnegie Publishing, 1992.

4   Power and the People

In the nineteenth century, industrialization picked up steam. Men adopted machines in more sectors of the economy and ran them in purpose-built structures using inanimate power. Factories accelerated the separation of production from consumption and work from home, while ships full of raw materials carried agricultural commodities across the seas to be processed by machinery in mills – by capital investments and the operating costs of labor. These practices had moved into view as part of textile industrialization. Steam power helped them expand. In 1800, James Watt's patent expired, and experiments with new engines and improvements on his separate condenser idea proliferated. After 1825, steam became the prime mover in Britain's economy, and industrialization shifted gears.[1] As global historian Prasannan Parthasarathi puts it, industrialization in the nineteenth century switched "from cotton to coal." Energy historian Andreas Malm describes the result as a "fossil economy" in which industrial capitalism and fossil fuels (mostly coal in those days, but oil too) are indissolubly linked. Malm describes industrial capitalism as "a socio-ecological structure, in which a certain economic process and a certain form of energy are welded together."[2] This chapter examines how the link between industrial capitalism and steam power was forged, amidst a welter of alternative possibilities.

In retrospect, steam power was a natural fit for industrial production, and the iconic case of textiles provides a fine example. As more firms chose steam, factories could rise anywhere cheap coal could reach. This meant that new entrants had more location choices – they were not limited to old grain mills or sawmills on waterfalls. Regular coal

---

[1] G. N. von Tunzelmann, *Steam Power and British Industrialization to 1860* (Oxford: Clarendon Press, 1978), 4–5; Joel Mokyr, *The Enlightened Economy: An Economic History of Britain 1700–1850* (New Haven: Yale University Press, 2012), 82.

[2] Prasannan Parthasarathi, *Why Europe Grew Rich and Asia Did Not: Global Economic Divergence, 1600–1850* (Cambridge and New York: Cambridge University Press, 2011), 151; Andreas Malm, *Fossil Capital: The Rise of Steam Power and the Roots of Global Warming* (London and New York: Verso, 2016), 12.

shipments could lower transportation costs in the new locations, so acquiring raw materials and distributing finished goods would also become cheaper than at older rural mills. Getting goods seen by buyers was also easier in the industrial towns, with lots of warehouses and transport links – including Manchester, now a marketing center for surrounding manufacturing villages. Because steam-powered mills could go more places, they could also be placed where workers seeking wages could be found. Watermills also had disadvantages. They sometimes slowed or stopped working due to drought or flood, delaying the return on investment they offered. Workers could also become restless, and "chafe at their loss" of hours and pay at these moments; some moved to town "for the benefit of regular work and wages which *steam* power ensures for them." After steam triumphed, it helped manufacturers control transaction costs, attract workers, and calculate regular returns on their investments.[3] These advantages were results, more than causes, of the new systems.

As steam power expanded, the technology of waterwheels also advanced. In Styal, outside Manchester, Samuel Greg still ran his Quarry Bank cotton mill with waterwheels. This did not make him backward, a straggler in adopting new machinery – quite the contrary. In the last decade of the eighteenth century, with profits from the French wars, he began to invest more capital into power supply. In 1796, he took as partner a man who had worked for the steam-engine firm of Boulton & Watt. Despite this fellow's experience with steam, he and Greg set out to improve Quarry Bank's waterpower system. By 1801, Greg had added two new waterwheels and built a stone weir and dam to control the flow of water in the system; six years later he added another wheel. By 1810, when he installed a Boulton & Watt engine, it was only to supplement his waterpower system.[4] His choices partly depended on the contingency of his site, but many firms in Manchester itself were making similar decisions about which power source best served their needs and resources.

For Greg, and many others, waterpower was clearly more efficient than steam – for those with the right location. Nineteenth-century overshot waterwheels could produce quite a lot of force – 300 horsepower. They were more efficient than steam engines, in the new meaning of the word that compared the relation of inputs to outputs. The best compound steam engines reached only 13.5 percent efficiency, even in

---

[3] "Factory Consolidation Act, Clause Excepting Mills driven by Water-Power," Royal Manchester Institution Papers, Manchester Archives and Local Studies, Manchester Central Library, Manchester, UK.

[4] Mary B. Rose, *The Gregs of Quarry Bank Mill: The Rise and Decline of a Family Firm, 1750–1914* (Cambridge and London: Cambridge University Press, 1986), 22–25.

the 1890s – and before the 1850s, less than four percent was common.[5] The figures represent how much of the energy locked in coal an engine could use, and waterwheels could put 85 percent of the power of falling water to industrial use. In addition, waterfalls cost less to operate as they required no coal. Waterpower would always be more popular than steam, a professor friend explained to James Watt in 1807, due to the cost of coal. A leading cotton manufacturer described water as "power for nothing, and in abundance." He was no small-scale operator, nor a technological laggard, any more than Greg was: he employed 2,300 workers in three factories. Nor was he mistaken. The River Irwell, on its approach to Manchester, fell for 900 feet of splashing waterfalls. As late as 1835, the river and its branches supported the operations of 300 individual mills.[6]

So how and why did steam win and the fossil economy take shape? It is not that steam-engine technology was superior to other options: the data we have discussed demonstrate that it was not, and this remained true through the mid-nineteenth century. Steam engines freed factories from water-based locations and produced other advantages that accelerated industrialization into more sectors. But the process was determined by politics, by social conflict, by the goals of the people who chose steam over waterpower, and the political forces that they marshaled behind their choices. Their selections depended on the problems they hoped to solve. Manchester's mill owners faced difficulties with their workers, who wanted more pay, and who also sought political power and status to improve their conditions. Manufacturers used steam to power new devices designed to replace recalcitrant workers, and they demonstrated along the way that capital investment helped make workers interchangeable, embodying their skills in the repetitive operations of machines. Observers described mechanization as the replacement of skilled artisanal labor with unskilled machine-minders, but as we have seen, the industrialization process had more stages and systems than that. By the end, though, commodified laborers recognized their condition in collective action, class identity, and political activism. This well-known story – in which technical change sparked worker resistance, which inspired further mechanization – now appears as one phase of the typical, complicated, back-and-forth relationship between social and technical change.[7]

[5] Astrid Kander, Paolo Malanima, and Paul Warde, *Power to the People: Energy in Europe over the Last Five Centuries* (Princeton and Oxford: Princeton University Press, 2013), 154, 182.

[6] Kirkman Finlay, testimony, Parliamentary Papers (1833) VI, 73; quoted in Malm, *Fossil Capital*, 86; Edward Baines, Jr., *History of the Cotton Manufacture in Great Britain* (London: H. Fisher, R. Fisher, and P. Jackson, 1835), 86n.

[7] E. P. Thompson, *The Making of the English Working Class* (London: Victor Gollancz, 1964); Trevor Griffiths, Philip A. Hunt, and Patrick K. O'Brien, "Inventive Activity in the

Technology Transfer: The New World

Comparing cases of industrialization shows how many different ways machines and methods could be arranged to facilitate mass production, and helps answer the question of causation posed at the start of this chapter: why did steam power work then, among the range of options? Cases of technology transfer help indicate the array of elements that go into making a machine work. The Arkwright system spread fast, not only to Scotland (remember Robert Owen's philanthropic mills in late eighteenth-century New Lanark) but also to the United States. New England possessed some of the features that had characterized preindustrial textile production systems in Britain. The region already had textile merchants, with capital invested in mills and machinery, usually jenny workshops spinning weft. Efforts to spin yarn strong enough for warp even resulted in attempts to power jennies by waterwheels, but these experiments failed (possibly by contingency – poor sites were chosen) leaving the jennies hand-powered as they were in England, even though waterpower was regularly applied to work the preparatory machines.[8] The array of choices, in American states that had recently been British, shows some of the many paths and outcomes of industrialization.

Samuel Slater (1768–1835) had been the manager of an English textile manufacturing business, an apprentice to Jedidiah Strutt (Derbyshire hosier and Arkwright investor) before he immigrated to the United States in 1789. His hands-on experience was invaluable to eager industrialists, and he was running a workshop of jenny spinners when the Browns of Rhode Island found him. The Brown family operated a worldwide trading firm with considerable experience dealing in slaves. They were also putting-out merchants in Rhode Island who placed wool and flax into the homes of pickers and spinners, did deals with spinners and weavers, and sold the finished cloth. They tried a weaving workshop but could not make it work, so instead they sold linen warps to home weavers. They later shifted to putting out their own yarn, and employed as many as 566 domestic weavers in the years between 1816 and 1820. With the help of Slater's direct experience of the Arkwright system, they first mechanized their spinning operations to serve this putting-out business: carding engines and roving machines, waterpowered frames for spinning, and children seven to twelve years old to operate them,

British Textile Industry, 1700–1800," *Journal of Economic History* 52 (Dec. 1992), 893; Gillian Cookson, *The Age of Machinery: Engineering the Industrial Revolution, 1770–1850* (Woodbridge: Boydell Press, 2018), 28.

[8] Gail Fowler Mohanty, *Labors and Laborers of the Loom: Mechanization and Handloom Weavers, 1780–1840* (New York and London: Routledge, 2006), 47, 49, 68.

complete in a purpose-built two-and-a-half-story factory by 1793. The mill employed a hundred people by the end of the century.[9]

By 1800, though, Samuel Slater had ventured out on his own. He created a different industrial system that borrowed some elements from the Arkwright model but also made some novel choices. He used water-power and his mills were located in rural settings. Partners supplied capital, typical of a mercantile business. In finding workers, however, he used a different formula. Instead of the apprentice system employed in the first British mills, or the free labor developing in Manchester, where workers found their own housing, Slater borrowed the patriarchal household from the domestic system of production. He hired entire families and paid the male heads of household for the whole family's work. Few of the men worked in the mill, however – 90 percent of the workforce was women and children. Men who did work in the mills were either super-visors or mule spinners, who managed and paid their sons as piecers. For the fathers without jobs, Slater built farms for them to cultivate. Sometimes their wives joined them, especially at harvest time. Weaving was done on a domestic system: as in Lancashire, Samuel Slater had children spin cotton into yarn on machines in factories and then he put the yarn out to patriarchal weavers in their homes. But the fathers in Slater's New England towns had the right to manage their children's work, even to enter the factories to supervise them directly, as well as to collect their pay. This Slater system spread across Rhode Island, Pennsylvania, and southern New England.[10]

Slater's mill villages looked a lot like the historic Puritan villages of New England. Linear town streets were bordered by open fields beyond, for men to farm. The church stood on Main Street, along with shops, stores, and maybe a post office, hotel, and tavern, with cottages for individual families. He left plenty of room for goats, chickens, and pigs, while larger farm animals, including horses and cattle, grazed on company land, as guild members and freemen had once had rights to use the town com-mons in England.[11] While this layout of buildings, factories, machines, firms, people, families, customs, and rights does resemble colonial New England villages, it also mirrored the plantation system that in the south-ern United States was being readily adapted to the agricultural

---

[9] Barbara M. Tucker, *Samuel Slater and the Origins of the American Textile Industry, 1790–1860* (Ithaca and London: Cornell University Press, 1984), 45–52, 71–75; Mohanty, *Labors and Laborers of the Loom*, 82–92.

[10] Tucker, *Samuel Slater*, 50–51, 85–86, 101, 110–111, 139–141, 223–25; David J. Jeremy, "Innovation in American Textile Technology during the Early Nineteenth Century," *Technology and Culture* 14, no. 1 (Jan. 1973), 44, 48, 74.

[11] Tucker, *Samuel Slater*, 124–29.

production of cotton fiber. There too the household formed the basis of production, albeit the enlarged household of which the master thought himself the father. There too most laborers lived on the property, but away from the master's house, in family units.[12] Patriarchy fueled the new industrial system in the one-time colonies – or rather, it fueled a case of industrialization. Alternatives existed.

While Napoleonic wars were still ripping up Europe, a different American merchant adopted British textile technology and adapted it to American conditions, with very different results. Francis Cabot Lowell was a powerful Boston merchant who, in 1813, established a different combination of business structure, machine operation, and labor organization for cloth production. His company in Waltham, Massachusetts, provided a model that others borrowed. As a result, Lowell-style mills spread across northern New England before 1830. The machines themselves differed, as did the production systems within which they operated. Lowell used not only spinning machines, he had also seen a powerloom in England and he made this his basis for mechanizing textile production. He wanted to mass produce the yarn and the cloth, both together under one roof. In England, experiments with powerlooms had not really worked. Handloom weavers were cheap workers. They produced flexible product lines, with manufacturers providing warps to get what customers wanted. In England, handloom weaving increased as a result of the mechanization of spinning in England, tripling in the four decades after 1795. In New England, though, Lowell mechanized weaving and used inanimate power sources. The result was a new production system and also a new business structure. Mechanizing weaving meant not only new machines but also an entirely different set of choices to keep them in operation.[13]

Lowell needed a lot of capital to realize this version, and used the corporate legal form to achieve it. Manchester start-ups had been cheap: people could rent room and turning, and buy machines from firms that went bust or moved or expanded. But, starting from

[12] Elizabeth Fox-Genovese, *Within the Plantation Household: Black and White Women of the Old South*, new ed. (Chapel Hill: University of North Carolina Press, 1988); Philip D. Curtin, *The Rise and Fall of the Plantation Complex: Essays in Atlantic History*, 2nd ed. (Cambridge and New York: Cambridge University Press, 1998).

[13] Jeremy, "Innovation in American Textile Technology," 45–59; George Unwin, *Samuel Oldknow and the Arkwrights* (New York: Augustus M. Kelley; 1968, orig. pub. Manchester University Press), 45–46; Peter Kriedte, "Decline of Proto-Industrialization, Pauperism, and the Sharpening of the Contrast between City and Countryside," in Peter Kriedte, Hans Medick, and Jürgen Schlumbohm, *Industrialization before Industrialization: Rural Industry in the Genesis of Capitalism* (Cambridge and London: Cambridge University Press and Paris, Editions de la Maison des Sciences de l'Homme, 1981), 156.

scratch, Lowell needed at least a third of a million dollars. Rather than find partners, he sought and received a charter of incorporation from the Commonwealth of Massachusetts. In the early nineteenth century, incorporation was a rare way to organize business, and only granted by special dispensation. As the Crown gave market rights in medieval England, and also granted corporate privileges to towns and guilds, so too was incorporation a special gift of the legislature of an individual state in America. A chartered corporate body could be divided into shares that could themselves be bought and sold, and the purchasers of these stocks were not responsible for the damage the corporation might do. Selling shares meant that a corporation could raise capital to invest in fixed costs like machinery or canals. A later firm on Lowell's model could require as much as two million dollars of capital investment before 1840, and incorporation made investment attractive. The scale of investment, of power and mechanization, meant specialization and mass production developed together. Each firm in the Lowell system specialized, weaving only one or two types of cloth. They did not compete with one another.[14]

Lowell's labor force was also new. Lowell and his imitators recruited workers for their factories among the unmarried women of New England and provided boarding houses around the mills for their lodging. These teenagers lived closely supervised lives under the eyes of "matrons of the highest respectability." Employment options for these women had previously been limited to teaching school, minding children, or domestic service. The higher wages of the mills, the more urban setting, and the adventure of living together outside the family made factory work an agreeable alternative for many of these workers, who came to factory towns from all over Massachusetts, New Hampshire, Vermont, and Maine. In 1820, out of a force of 264 workers, 225 of them were women and girls, 13 were boys, and 26 were grown men. Despite the significant technological differences between Slater's and Lowell's textile mills, in both cases grown men formed no more than 10 percent of the workforce (not including male mechanics and bleachers). In Lowell's mill villages, young women lived in boarding houses. Nonetheless, the mill village took on the form of both a household where people lived, and also a production complex, not so unlike the plantation system of the American South. Its buildings and layout also resembled the Slater-style factory towns that combined family farms and cottage

---

[14] Tucker, *Samuel Slater*, 113–15; Jeremy, "Innovation in American Textile Technology," 45–46.

households, but without the individual, father-headed households on which Slater relied.[15]

In Philadelphia, textile manufacturing took yet a different form. There, the ingrain carpet manufacturers used factory production to produce small batches of goods that changed readily to meet the demands of fashion. Their production systems reflected a blend of industrial and artisanal orders. Individual workers claimed specific machines as their own to operate. The manufacturers were proprietors, who treated their business enterprise as an extension of themselves. Many retired out of the industry once they had earned enough money; without corporations, businesses rarely outlasted their owners. Philip Scranton has described this form of industrialization as proprietary capitalism to distinguish it from the corporate business form used by Lowell and his imitators in Massachusetts. Scranton articulated a "matrix of accumulation" in which production and organizational elements, machines and factories and power sources, workers and raw materials and designs, were selected from an array of possibilities. This was true across the industrialization experience, from eighteenth-century Lancashire to the United States, and across its several structural variants.[16]

Technology does not transfer on its own, and individual machines do not explain how production systems develop. Transferring technology is not as simple as moving a machine to a new location, just as invention is more complicated than an individual genius experiencing a flash of discovery. McConnell & Kennedy knew that – as they adapted machines to the needs of individual customers, and when they recommended mechanical and organizational choices based on their experiences and those of their clients. Even James Watt sent individual engineers to erect steam engines and make the necessary changes and counsels on site. Samuel Slater's English experience made him invaluable in transferring the Arkwright system to the US, but the system changed as it was adapted to the social structures of New England and adopted its precepts in order to operate. Historians of science and technology often classify the need for humans to physically accompany moving machines in terms of tacit knowledge – known and conveyed in person rather than through printed sources or artifacts. But even with people minding the set-up and

[15] Jeremy, "Innovation in American Textile Technology," 46; Tucker, *Samuel Slater*, 113–115.

[16] David J. Jeremy, "British Textile Technology Transmission to the United States: The Philadelphia Region Experience, 1770–1820," *Business History Review* 47, no. 1 (Spring 1973): 24–52; Philip Scranton, *Proprietary Capitalism: The Textile Manufacture at Philadelphia, 1800–1885* (Cambridge and New York: Cambridge University Press, 1983).

arrangement of operations, still the array of choices in machines, capital sources, investment and business organization structures, and labor and power systems, reveals how complex is the process of industrialization and technological change.

Industrializing Traditional Textiles

New technological systems for making other kinds of cloth were emerging in Britain almost as rapidly as those for cotton. Of course, John Lombe had borrowed Italian technology and successfully mechanized silk throwing in Derbyshire in the first part of the eighteenth century – an early model for cotton industrialization – and, later in the century, John Marshall introduced machines and power to turn flax into linen across the river from Leeds. Woolens and worsted have also provided examples of the intermittent mechanization of the trade, and how consumers at some distance from the factory received goods through distributors. But certain episodes of industrialization in these textiles have been woven into the mythological history of the cotton industry of Lancashire – the Luddites, for example, have often provided simple lessons about resistance to the Industrial Revolution. But they had little to do with the mechanization of spinning or weaving, nor with the cotton industry of Lancashire. Instead they function as part of the myth, and at the same time show the more complicated history of technological change than the myth can capture. The original Luddites of legend and lesson were mainly wool finishers of Yorkshire, or displaced hosiers and knitters from the Midlands. Their story mingles industrialization into tradition, and shows how older ways were shaping the new technology.

After all, cotton and wool were not entirely separate businesses. Many of the earliest machines associated with the Industrial Revolution – especially the pre-Arkwright machines, including John Kay's flying shuttle – were intended for use on fleece fibers before their successful versions were applied to cotton. In other words, machines designed for wool were often not adopted, but made to work on cotton instead. However, a machine for combing wool was patented in 1789 – while cotton spinning was still developing into factories. And some merchant-manufacturers made more than one fiber into yarn or cloth, despite different supply chains – the Browns who partnered with Samuel Slater in the United States did different businesses in wool, linen, and cotton.[17] In Lancashire itself, the Helmshore Mills contained facilities for carding and spinning yarn from fleece, for fulling wool pieces, and also for spinning cotton, in several

[17] Griffiths, Hunt, and O'Brien, "Inventive Activity," 889; Tucker, *Samuel Slater*, 50, 99.

buildings on one site. The firm sometimes worked in cotton and sometimes in wool, throughout the nineteenth century. As Samuel Oldknow combined factory spinning and handloom weaving of cotton, other men had similar operations, but in more than one fiber. Henry Sagar, for example, was a cotton manufacturer with a spinning mill but also a worsted comb-shop and a warehouse to manage his putting-out business. The roots of industrialization in domestic and putting-out businesses meant that early industrialists did a range of business in what historians consider separate sectors.[18]

Nonetheless, each branch had its own history, and the business structures of the woolen and worsted industries resulting from those histories partly account for their differing adventures in industrialization. Merchant-manufacturers of the New Draperies often invested more capital in raw materials, dividing up tasks and putting their worsted bits out to households performing separate production tasks. The old difficulties of the putting-out system did persist: in 1817, a merchant traveled 762 miles across Yorkshire in search of worsted yarns. He chased down yarn at markets and cloth halls across the region. In Bradford, he remarked the appearance at the Cloth Hall of excellent machine-spun worsted yarns, and described the mills and manufacturing operations that produced them. Preindustrial textile networks accommodated the introduction of machinery, and the new products supported the old institutions of exchange. In the nineteenth century, as machines for preparing and spinning wool and worsted yarns began to work, they were adopted in ways that could fit into existing patterns of business organization.[19]

While worsted, the New Draperies, had grown up in Yorkshire in proto-industrial forms, wool often stayed in more old-fashioned patterns. Fulling had already left the home in the medieval period, but most wool in eighteenth-century Yorkshire came from individual homes. There men still wove a piece or two at a time, to sell some weeks at the public market, or to a merchant or clothier, who usually arranged for further finishing in a nearby textile town. Early-modern gig-mills had long threatened the finishing trades – these devices could raise the nap by machine for cropping down into a smooth fabric. Laws had suppressed the adoption of these machines in the early 1550s, but not entirely successfully, and the

[18] Cookson, *Age of Machinery,* 20–21, 38–39; Christopher Aspin, *The Water-Spinners* (Helmshore: Helmshore Local History Society, 2003), 222; Helmshore Mills Museum, Rossendale, UK (visited in summer 2013).

[19] Pat Hudson, *The Genesis of Industrial Capital: A Study of the West Riding Wool Textile Industry, c. 1750–1850* (Cambridge: Cambridge University Press, 1986), 29; Gillian Cookson, "A City in Search of Yarn: The Journal of Edward Taylor of Norwich, 1817," *Textile History* 37, no. 1 (May 2006), 38, 43, 48.

Crown prohibited them again, more thoroughly, a hundred years later. In 1737, specialist wool finishers from Lancashire were still complaining that the machinery was going to replace them.[20] But in the nineteenth century, such sporadic mechanization in the woolens business quickened into industrialization. People no longer made their own clothes, not the ones for show; they bought what they wore with money earned by working for others. Raw materials no longer grazed on the weavers' land, awaiting the annual shearing. Even fifty years earlier, the system had been very different. Now machines for preparing and spinning the fleece were being adopted piece by piece, hooked up to run on inanimate power, as merchants sunk their money into machinery, and their connections into supply and distribution networks. They became manufacturers in the process.

Benjamin Gott (1762–1840) was a wool merchant who built a factory just outside Leeds. He stayed a merchant, even as he integrated all the processes of wool production, from preparation through weaving to finishing, under one roof. He had built his Bean Ing Mill between 1792 and 1794 – earlier than Lowell would do for cotton in the United States. He used a Boulton & Watt steam engine from the start, and utilized it for those processes that others drove with waterwheels: preparation, carding, and fulling, and also for grinding dyes. However, most of the work in his factory was carried out by hand-power. All the weaving was done on handlooms. Gott employed 761 workers in 1813, and more than a thousand in 1819: in that year, three-quarters of his workers were still engaged in hand processes.[21] While Gott masterminded the production line, he sub-contracted the manufacturing processes to five tenants who worked independently, producing cloth to order for Gott's commercial firm. By 1807, Gott had shifted operations into Armley Mills, an even more traditional factory just outside Leeds, waterpowered, that provided public access to his preparatory and fulling machinery. Domestic weavers, workshop masters, and merchant clothiers took their goods to his mill and paid for processing, then returned home with wool for spinning, or cloth to be finished. Gott's woolen business at Leeds demonstrates the

---

[20] Malcolm I. Thomis, *The Luddites: Machine Breaking in Regency England* (Newton Abbot: David & Charles, 1970), 15; John Smail, *Merchants, Markets and Manufacture: The English Wool Textile Industry in the Eighteenth Century* (Basingstoke and New York: Macmillan and St. Martin's Press, 1999), 69, 140; John S. Lee, *The Medieval Clothier* (Woodbridge: Boydell Press, 2018), 59, 168.

[21] R. G. Wilson, "Gott, Benjamin (1762–1840)," in *Oxford Dictionary of National Biography* (Oxford: Oxford University Press, 2004); Wilson, *Gentlemen Merchants: The Merchant Community in Leeds, 1700–1830* (New York: Manchester University Press, 1971), 243, 248; Herbert Heaton, "Benjamin Gott and the Industrial Revolution in Yorkshire," *Economic History Review* 3, no. 1 (Jan. 1931), 45–61.

persistence of ordinary production systems even as they were consolidated under one roof using the latest machinery.[22]

This combination of all technical procedures within a single firm would later become known as vertical integration. From bottom to top, raw materials to finished product, all operations worked together. The enduring use of handiwork, even inside vertical mills, demonstrates that machines were only part of wool's industrialization. Mechanization was the result of many people choosing specific methods and machines within the structures already used in industry, usually changing those institutions as they made it all work. Established merchants understood the latest developments in other textile branches and adopted what worked for them, including factories and steam power for some processes, hand and even domestic production for others: as in cotton, some home-based woolen and worsted workers received a boost from the industrializing system. While Gott brought these workers under his roof, other firms stayed with more routine putting-out systems. Design and distribution also changed to respond to the accelerated fashion shifts of the nineteenth century.[23] Manufacturing developed hand in glove with new organizations of the merchant business, as it did in cotton, and both sectors drew on familiar forms.

A pair of conflicts between Benjamin Gott and the finishers in his factories shows how industrialization incorporated historic structures even as it transformed or destroyed them. One case concerned apprenticeship. Gott had come up the old way, from apprentice to partner before controlling the firm. In 1802, he took on two new apprentices in cloth-dressing – in the finishing division. The young men were over fourteen years old and the journeymen refused to work alongside boys who would be free from apprenticeship (at age 21, as was customary) before completing seven years of training. The journeymen based their refusal on the Tudor apprenticeship laws from the 1560s, and Gott replied with the Combination Acts that only two years earlier had outlawed trade unions and collective bargaining. Gott tried putting the work out to other merchants, but all the finishing men in Leeds refused to touch the cloth. Gott agreed to a legal arbitration in order to obtain a quick resolution. The result, agreed to by all the masters and all the men in the Leeds finishing trade, reaffirmed the Tudor laws and the rights of journeymen.[24]

---

[22] Cookson, *Age of Machinery*, 23; Colum P. Giles and Ian H. Goodall, *Yorkshire Textile Mills: the Buildings of the Yorkshire Textile Industry, 1770–1930* (London: Royal Commission on Historical Monuments, 1992), 85–87.

[23] Smail, *Merchants, Markets and Manufacture*, chapter 7.

[24] Heaton, "Benjamin Gott," 58–59.

Apprenticeship also persisted in the most innovative industries, including the engineering firms building steam engines and textile machinery. The new engineering developed out of the established handicraft trades, learned under long contracts and spread in the tramping of journeymen among forges and foundries. But in more familiar businesses, the institution was fading. Guilds were losing members and control of their trades, and their roles in local government. Historians argue about the reasons for this decay. Some think that the guilds' monopoly privileges, and the limitations they placed on access to markets, demonstrate their economic inefficiency. Others argue that guilds were efficient institutions that preserved and passed on craft knowledge and that smoothed trade through reputation mechanisms and through the regulation of supply and demand. But the most detailed analysis of the archives left by English guilds argues that the increasing role of export markets divided guilds internally. Large overseas traders had interests that differed from those of smaller merchants, who stayed within local or regional networks. These differences pulled members apart, and damaged the internal cohesion of the groups. Guilds collapsed on their own, even while industrialization was happening. In this case, the law was catching up to changes already underway: in 1814, Parliament repealed the Tudor laws that had established the terms and obligations of the master-apprenticeship relationship.[25]

Benjamin Gott's other clash with the wool finishers concerned the gig-mill. In the last years of the last century, Gott had tried to introduce a gig-mill at Bean Ing. He received threats from machine-breakers, and headed up a private police force intended to protect his devices, but the Mayor of Leeds persuaded Gott and the merchants to stand down. They gave up the gig-mill. The machine had existed for centuries, and versions had been patented in the 1780s and 1790s. The basic idea entailed rotating large frames of teazels (dried thistles) against the fulled wool, which raised the nap for cropping. These machines had been employed on a small scale in Yorkshire in the early 1790s, without incident. But the number of gig-mills in Yorkshire increased from five to seventy-two in the single decade after 1806. Manufacturers were using steam to power the big shears that cut the nap, which increased their incentives to use those same engines to lift up the nap to the shears. The number of machine

---

[25] Cookson, *Age of Machinery*, 153, 160; Regina Grafe, "Review of Epstein and Prak, *Guilds, Innovation, and the European Economy*," in *Journal of Interdisciplinary History* 40, no. 1 (Summer 2009): 78–82; Michael John Walker, "The Extent of the Guild Control of Trades in England, c. 1660–1820; A Study Based on a Sample of Provincial Towns and London Companies," (Ph.D. Diss., University of Cambridge, 1985), 218, 219.

Figure 4.1 This view of a reconstructed gig-mill shows the iron frames partially filled with teasels (dried thistles), which raised the nap on a piece of wool cloth. The gig-mill was one of the machines singled out for breaking by the Yorkshire Luddites in the 1810s. Photographed by the author in 2015 at Leeds Industrial Museum at Armley Mills, Yorkshire.

shears in Yorkshire grew from 100 to 1,462 between 1806 and 1816, the same rapid rate of adoption as the gig-mills. When the two were made to work together using inanimate power, two thousand men were briskly put out of work. The new use of machinery that had existed for centuries led to immediate unemployment. The reaction was violent.[26]

Luddites

Wool croppers made redundant by the steam-powered technological package of finishing processes around the gig-mill resorted to traditional remedies. They rampaged across the industrial landscape, destroying textile machinery, between 1811 and 1817. Named for their mythological leader, the Ned Ludd or General Ludd who signed letters threatening mechanizers, Luddites by definition conspired to attack industrial machinery. The movement associated with the name began in the Midlands, in the Nottingham lace and hosiery business. In 1811, a "villainous reputation" was assigned (first by journalists, later by historians) to the industry's knitting frames. Reports spread that this invention would make thousands of workers obsolete. The machine, of course, already existed: it had been in common use for two hundred years. But as machines were adopted into more rooted production systems, Luddites broke into workshops and smashed knitting frames across the Midlands through 1816. In Yorkshire, Luddism among the displaced wool croppers was even more savage, and so was the response. Men were killed by gunfire on both sides in several separate battles. The Luddites were eventually caught and tried, which provided moral lessons about breaking the law and the power of capital to punish opposition. Seventeen Yorkshire Luddites were found guilty and eventually executed at York.[27]

The Luddites have become mythological examples of resistance to technology, almost a generic description, a cultural category. But of course each action assigned that name also had its own specific context. Many incidents embodied anger toward a singularly unpopular master: the murder of William Horsfall, wool merchant of Yorkshire, counts in this category. But the Luddite machine-breakers were also part of several longer movements emerging from the region's history of radicalism, and becoming political activism, class identity, and the quest for Parliamentary representation. Hunger was also stalking the region in the 1810s: bad harvests and high food prices were making people desperate. Machine-breaking was one of the customary ways that working

---

[26] Cookson, *Age of Machinery*, 192–93; Heaton, "Benjamin Gott," 58–59; Thomis, *Luddites*, 11, 29, 50, 165–85.

[27] Thomis, *Luddites*, 29, 177–85; Cookson, *Age of Machinery*, 192–93; Jeff Horn, "Machine-breaking in England and France during the Age of Revolution," *Labour* 55 (Spring 2005), 146–47.

people had expressed their prerogatives, going back to the guilds. In 1710, a London guildsman had his knitting frames broken when he took on more apprentices (forty-nine of them) than the Worshipful Company of Framework Knitters permitted. When knitters and croppers adopted similar methods in Yorkshire a hundred years later, radicalism and religion still shaped their actions. The region's networks of seething unrest included trade unionists, parliamentary reformers, and food rioters. Authorities called them Jacobins, a term that recognized the political nature of the outbursts. A Leeds newspaper compared the Luddites to thieves and robbers. Elites believed that working people were becoming lawless criminals in defense of their interests.[28]

Nineteenth-century Luddites performed vicious, illegal acts in response to the stresses wrought by rapid and revolutionary technological change, but their actions also had fundamental causes in hunger and economic misery, with deeper roots than the direct threats to their employment. They applied customary responses to their immediate conditions. They were part of a wider movement of workers displaced when steam power crushed the older artisanal structures on which it was being erected. The calico printers, for example, also transformed old ways of expressing anger into political activism. The textile printing industry had first sprung up in England in the late seventeenth century in response to Indian imports, under Parliamentary protection. It had sprouted under protective cover and it used older guild-type structures. In the first decades of the nineteenth century, the old trade dissolved in familiar ways. Journeymen printers fought against the masters for adding excessive numbers of apprentices to the trade. The journeymen struck for months in 1815, but found themselves readily replaced by new machinery. Displaced calico printers sought radical solutions and, in the next decade, joined the Chartist party, full of workmen seeking political representation for the laborers of the new industrial order. They were not called Luddites, but their hatred of machinery was legendary.[29]

Manchester

The misery of working families was not limited to the Luddites who roamed the countryside mills of Yorkshire and the Midlands, breaking

---

[28] Thomis, *Luddites*, 13–27, 45; Malcolm Chase, *1820: Disorder and Stability in the United Kingdom* (Manchester and New York: Manchester University Press, 2013), 5–6; Eric J. Hobsbawm, "The Machine Breakers," *Past and Present* 1 (Feb. 1952), 58; Thompson, *Making of the English Working Class*, 521–22, 569–73.

[29] Geoffrey Turnbull, *A History of the Calico Printing Industry of Great Britain*, ed. John G. Turnbull (Altrincham: John Sherratt and Son, 1951), 182–95.

machinery and murdering masters. In Manchester, too, the brutal nine-teenth-century workplaces nursed resistance. Machine operatives in 1825 worked between twelve and fourteen hours a day without a glimpse out of doors. Men, women, and children all worked in the factories. In 1833, the cotton mills employed 60,000 men and 65,000 adult women, and 43,000 boys and 41,000 girls under eighteen years of age. About half the children were younger than fourteen.[30]

Discipline was rigid: a spinner could be fined for being late or sick or whistling, for being dirty but also for washing himself. He was fined when he was absent if he could supply no replacement: he was charged for the steam spent on his mule with no operator to run it. Spinners were paid by how much yarn their machines spun. They paid their piecers (boys who twisted together any broken threads) out of their own wages. Factories were hot and poorly ventilated; the greasy machines and the cotton fiber filled the noisy rooms with moist, cotton-dusty air that was difficult to breathe, that coated the big windows and the iron machinery, sticky from dirty oil. Gaslights, grease, and cotton lint combined to make fire a serious hazard. The clanking, lumbering mules were powered by turning over-head shafts. Long leather belts snaked down to each individual machine, spinning incessantly from the giant rods that turned with the heavy motion of the steam engines or waterwheels. Mill operatives worked in a hot, wet, lint-filled space filled with heavy moving machinery that made a constant, deafening, racket.[31]

Housing in Manchester was also famous for its horrors. An 1808 visitor described a smelly, filthy city that stank from coal-fired steam engines and dyehouses, with river waters that seemed curdled, black and thick with oil and refuse. It was only getting worse. In the 1840s, Friedrich Engels studied the conditions of the English working class, in its city of origin. His depictions would form one basis for Karl Marx's famous critique of capitalism. In the oldest part of Manchester, the crowded and narrow Todd Street, Withy Grove, and Shude Hill streets twisted between the commercial district at Market Street and the Irk River, itself a "coal-black, foul-smelling stream." Full of garbage, the river supplied the only water available to the houses. Behind the main streets, covered passages between buildings led further and deeper into networks of squalid court-yards, surrounded entirely by tenements, where one privy often served all the surrounding inhabitants. Some residents kept pigs in these

---

[30] J. L. Hammond and Barbara Hammond, *The Town Labourer, 1760–1832: The New Civilisation*, 4th impression (London: Longmans, Green, and Co., 1919), 23n2.

[31] Hammond and Hammond, *Town Labourer*, 19–20, 164; Robert G. Hall, "Tyranny, Work and Politics: The 1818 Strike Wave in the English Cotton District," *International Review of Social History* 34, no. 3 (1989), 450.

courtyards – a source of food, but also of foul waste. People lived in cellar rooms that accumulated pools of excrement on the floor. Rows of houses built back-to-back prevented air from circulating, while unpaved streets without sewers flowed with filth. In the nineteenth century, Manchester was famous for its misery (See Figure 3.3).[32]

These true stories of pain and squalor have achieved mythological status, partly because they form an effective backdrop for a triumphant tale of political reform and sanitary regulation in the nineteenth century. In 1832, Parliament would expand the franchise and grant some property-less (male) workers the right to vote. The campaign to repeal the Corn Laws in the next decade expressed for the Manchester manufacturers an ideology and identity that brought them political power. The partisan effects of the formation of class identity should not, however, obscure the increasing opportunities and hopes presented by the new industrial order. Forbidden by law from collective action, without political voice or representation, Manchester's workers were not without opportunities for self-expression. A surprising number of working-class men wrote down their autobiographies in the 1820s and 1830s, some for publication and some for the edification of family and future generations. Most of these sketched life stories in which their origins in poverty gave way to opportunity – between a third and a half were writing the tales of their success in business or politics, arts or religion.[33] It is also clear that work for wages gave many Britons the hope and expectation of more: more clothes and more freedom, more pay, plus the vote, eventually, even, a system of government that better reflected and represented the changing structure of society.

Workers were employing familiar tactics and resources both to protest their condition and demand change. In summer 1818, against a backdrop of political disaffection, outdoor meetings, and speakers who advocated parliamentary reform and universal suffrage, a wave of strikes swept Lancashire. Artisans and jenny spinners, handloom weavers and factory workers, hatters and tailors and coal miners, turned out of their workplaces for higher wages in towns and villages across the North. The climactic action that summer was a giant strike of Manchester's cotton spinners. Ten thousand of them quit their workplaces in July, demanding more pay. As the strike spread, as many as 30,000 workers may have

[32] Frederick Engels, *The Condition of the Working Class in England in 1844, Complete and Uncensored*, translated by Florence Kelley Wischnewetzky ([Springfield, MA]: Seven Treasures Publications, 2009; orig. pub. 1844), 42–46; Malcolm I. Thomis, *The Town Labourer and the Industrial Revolution* (London: B. T. Batsford, 1974), 52.

[33] Emma Griffin, *Liberty's Dawn: A People's History of the Industrial Revolution* (New Haven: Yale University Press, 2013).

turned out of Manchester's mills. They picketed regularly throughout the city: several thousand would assemble and march through the streets, three or four abreast, bearing placards and banners expressing their grievances. Intimidating in their ordered masses, the marching strikers inspired workers in the other trades. Rumors of a general "Union of Operative Workmen, Mechanics, and Artizans of the United Kingdom" threatened to bring all laborers together to stop the economy. The 1818 Lancashire cotton spinners' strike seemed to menace the entire industrial order.[34]

Then, in late August, the mill owners re-opened their gates and restarted production – using new operatives. Rather than negotiate with the striking spinners, capital meant to replace them. The strikers responded with force. They would appear at the gates of the mills at the start of the workday to heckle and threaten their surrogates. Women and children at the fore, every day the mob would quickly gather and just as quickly disperse, harassing the spinners the masters intended to cover the shift. Eventually magistrates sent armed men to meet the strikers in the streets. On September 1st, the picketing spinners of Manchester marched to Stockport to support the rest of the trades in their work stoppages. By then, however, their strike was already collapsing into street fights. When they showed up at factories to stymie the new hires, gunfire from the rooftops scattered them. Their pickets were called mobs that surrounded "respectable gentlemen" to hoot and hurl insults.[35] The strike failed. The strikers dispersed, and the machinery kept spinning. Class struggle was not an abstract condition of industrial capitalism. It took shape in real violence. The Luddites roamed the wool districts, the knitting towns. The filthy streets of Cottonopolis were filled with unemployed picketers who marched in formation, who streamed to nearby villages to support their fellow strikers. Hunger, wages, anger, and the hope of political representation were weaving together an emerging working-class identity.

Peterloo

The existing social fabric was fraying as its threads wove together into new patterns. The next summer, it ripped wide open when a public meeting in Manchester devolved into a massacre. The gathering on 16 August 1819 was supposed to be a peaceful protest rather than a furious battle in which local police on horseback mowed down people fleeing on foot. The crowd expected a lovely day of public assembly and political oratory. They

---

[34] Hall, "Tyranny, Work and Politics," 433, 445–52, 458–59, quotation at 456.
[35] Ibid., quotation at 458.

dressed in their best clothes, and looked forward to eating outdoors in the scant Manchester sun. Families and protestors began arriving before noon at an empty field outside of St. Peter's Church on Peter Street. Some had brought picnics and some carried banners advocating universal suffrage, the reform of Parliament and whom it represented. The featured speaker of the day was Henry Hunt, a well-known reformer and friend of the people, but a man with an arrest warrant in his name.[36] His appearance at the gathering was an invitation to the police, who wanted him taken into custody.

The crowd swelled to somewhere between 80,000 and 100,000. Bugles sounded and a parade took shape. Behind a white silk banner, more than 150 women, members of a female reform group, marched in step. When Hunt climbed to the stage to speak, a local militia, mounted on horseback – merchants and manufacturers, pub-owners and shopkeepers – raised swords above their heads. They charged their horses through the masses to arrest him. People fell back from the assault, but sixteen were killed by sword. The crowd fled the slashing blades, trampling hundreds more when they fell. Some of the swordsmen penned in any stragglers and cut them down where they stood. The brutality included local men settling scores, but the national government congratulated Manchester magistrates on the successful arrest. Parliament responded with fresh crackdowns on any political speech or meeting, on parades with banners, or people marching in formation. Houses could be searched without a warrant. Any that charged admission for speeches or lectures were classified as disorderly – as houses of prostitution. The message contained in the massacre and its aftermath was contained in the name given the events. Peterloo marked the end of comity between capital and labor. It showed working people that both the local and national governments served capital, not labor – not themselves and their interests.[37]

For many historians, the Peterloo massacre of 1819 marks the beginning of working-class consciousness, one of the main effects of industrialization on social structure. The meeting was political. It was not an episode of machine-breaking or a strike for higher wages. It occurred outside the workplace and its speakers demanded political representation for workers without property, without capital. After Peterloo, according

[36] E. P. Thompson, *The Making of the English Working Class* (London: Victor Gollancz, 1964), 622, 683–90; Malcolm Chase, *Chartism: A New History* (Manchester and New York: Manchester University Press, 2007), 12–13; John Styles, *The Dress of the People: Everyday Fashion in Eighteenth-Century England* (New Haven and London: Yale University Press, 2007), 326.

[37] Donald Read, *Peterloo: The Massacre and Its Background* (Manchester: Manchester University Press, 1958), Thompson, *Making of the English Working Class*, 686–689; Chase, *1820*, pp. 6, 44–45.

to this argument, a new class structure had appeared from the feudal world of mutual obligation that appeared in patriarchal households, master–apprentice relations, and guild charters. When Parliament repealed the Tudor apprenticeship laws in 1814, the institutional basis for the preindustrial production system lost its legal footing. But working-class consciousness was emerging. Interchangeable workers felt camaraderie. Sparked by contingent local events like the Peterloo massacre, woven over time from familiar practices and imperatives, workers formed new institutions. In 1824, ten years after the Tudor apprentice regulations were repealed, Parliament recognized the new order by repealing the Combination Acts. This deregulation made trade unions legal and permitted workers to join together to negotiate the terms of their employment.[38]

Capital and Labor

The years around 1825 mark a turning point. Socioeconomic distinctions between capital and labor came into focus after Peterloo with the legalization of trade unions and businessmen's organizations. It was also in the second quarter of the nineteenth century that coal-burning steam engines came to dominate the productive economy of Britain. As in the expansion and condensation cycles of the steam in the engines, however, capital and labor worked both together and in opposition to each other, both part of the mechanism that made industrialization run. For example, strikes by workers could hurt manufacturers, but they could also be useful. When demand for his finished goods dropped, a manufacturer might cut wages to save costs. If workers in response went on strike, he saved their wages – plus the costs of coal to run the machinery. Idle machinery wasted capital, and interest bills came due, and no one could lose money that way and last for very long. But the savings created by striking workers could carry a lucky mill-owner through a slow period. With the capital sunk in machinery, cutbacks made elsewhere – in operating costs – could help.

Unions and collective activity by industrial workers are often painted as antithetical to the interests of capital, but they also embody and reflect the industrialization process. Preindustrial rituals for expressing grievances, machine-breaking and disorderly stoppages, were not only illegal, they were also unmanageable. It was hard for a mill-owner to identify a leader of activities deemed criminal – and difficult too to discuss the grievance, come to a solution, and get back to work. It is partly for this reason that

---

[38] Chase, *1820*, p. 4; Malm, *Fossil Capital*, 61.

Marxist historian Eric Hobsbawm calls preindustrial methods of expressing dissatisfaction, like the machine breaking of the Luddites, "collective bargaining by riot." Legalizing collective bargaining by trade unions made working-class identity real, and created an institutional basis for its political voice. The older traditions of workshops and guilds were changing to meet industrial needs. Most workers still did their jobs in older businesses, even as these expanded. But in the nineteenth century, according to economic historian Maxine Berg, even time-worn business forms operated "within the circuit of industrial capital," a new economic cycle of boom and bust that burst into financial crisis in 1825. These "waves of industrial credit creation culminating in periodic financial panics" that "rippled outwards" also stimulated "rationalisation of production and the standardisation of the product" across the economy. Expansion and contraction were spreading the industrial idea of mass production.[39]

Boom and Bust

When Parliament legalized trade unions by repealing the Combination Acts in 1824, the manufacturers were also battling other woes. In 1825, the financial system of Britain underwent a severe contraction. Its timing has marked it as the first real downturn of the modern business cycle. Caused by the financial system itself – which had grown during industrialization – and not by war or exogenous shocks, the contraction followed inflationary overexpansion of the circulating currency. The causes of both the boom and the bust stood outside the textile industry, but the credit requirements of industrial capitalism – the investment in machinery that waits and works for its return, preferably with interest – exposed the cotton manufacturers to considerable risk. The long boom in their business went bust in 1825.[40]

Britain had been at war with France for decades, punctuated by brief periods of peace. After 1805, during the Napoleonic Wars, high taxes had flowed to the Treasury, and the Bank of England had profited from each transfer of government funds to wartime suppliers. The Bank was issuing more banknotes and lending more money to London businesses, as well as to merchants who were moving their affairs from the unstable European Continent to London. City banks in London and country

[39] Hobsbawm, "Machine Breakers," 59; Maxine Berg, *The Machinery Question and the Making of Political Economy, 1815–1848*, rev. ed. (Cambridge and New York: Cambridge University Press, 1980), 1.

[40] Larry Neal, "The Financial Crisis of 1825 and the Restructuring of the British Financial System," *Federal Reserve Bank of St. Louis Review*, May/June 1998.

banks likewise grew in number and issue. Peace in 1815 did not stem the exuberance, but, in 1821, Parliament forced the Bank to resume converting its banknotes into coin – to resume the gold standard – putting hard money into circulation. But the Treasury continued to buy back its own bonds, which added to liquidity. Meanwhile, country banks were circulating small-denomination banknotes that aided manufacturers and kept production flowing. The easy money made lenders reckless and investors bold; the stock market reached a peak in April 1825, and country banks began to fail, continuing through autumn. Two London banks failed in December and the crisis reached the Bank of England over the Christmas holiday. Though it managed to keep making payments, the rest of the banks were not so lucky. Of 770 banks in England, 73 collapsed; so did three of 36 in Scotland. Textile manufacturers saw profits of 50 percent from the Arkwright era plunge below 5 percent. Bankruptcies swept the Isles, peaking in April 1826. Mills closed. Investment stood idle.[41]

The summer of 1826 was hot and dry – a drought disaster. The lack of rainfall inspired some mill-owners to switch to steam when low water prevented their waterwheels from turning. Even deer were dying of thirst, and the grass crop failed, leaving no hay for cattle for the winter. Oats rose to double the usual price.[42] The Corn Laws that had been imposed at the end of the French Wars were convenient scapegoats for the rise in food prices. These tariffs, a holdover from mercantilism intended to protect domestic agriculture, limited the importation of grains, and levied tariffs that raised their prices. The Corn Laws preserved income for landowners while the high price of bread punched a hole in workers' pay. High food prices and a government intent on keeping them that way; hot weather, failing firms, tight money: times were tough in the trough cycle of the industrial order.

Iron Men

Lancashire spinners revolted again in 1826, and turned out of the factories. They wanted higher wages to pay for unaffordable food. This time the mill owners turned to machines to replace them. They had long dreamt of a better mule, just a bit more automatic – especially the putting-up process, when the long iron carriage of spindles had to be returned to its starting point. This was the heavy part of the cycle, the steady shove from the whole strong back that made factory spinning a man's job.

---

[41] Ibid., 55–65; Malm, *Fossil Capital*, 58–61.

[42] Archibald Alison, *History of Europe, from the Fall of Napoleon to the Accession of Louis Napoleon* (Edinburgh: William Blackwood and Sons, 1855), 4:81; Malm, *Fossil Capital*, 168–70.

Figure 4.2 Spinning mules like this one were renamed "mule jennies," a reference to the unpowered spinning jenny, after the Iron Man headstock (attached to a machine like this one) made moving the carriage, mounted here with empty spindles, automatic. The rollers at the back recall those of Richard Arkwright's water-frame (see Figure 2.4). Courtesy of duncan1890/DigitalVision Vectors/Getty Images.

Manufacturers wanted fewer men, cheaper women, and more docile children to mind the machines.[43] They turned to Richard H. Roberts (1789–1864), a local mechanic known for an array of innovations – especially the ingenious machine tools that allowed smiths to cut gears precisely into iron wheels. A few years earlier, Roberts had already patented a device to automate the spinning mule, actually a headstock that could be fitted to existing, Crompton-style mules, to automatically put up the carriage. The manufacturers asked him to get this device working. It took until 1830. Once it got running, the self-acting mule threatened the whole class of mule spinners who operated existing machinery.[44]

Richard Roberts' headstock was intended from the start to be operated by steam power, whose regularity would operate the mule without stopping and starting, and make the movement of the mule a product of the engine. The self-acting mule, as mules outfitted with the headstock

[43] Mary Freifeld, "Technological Change and the 'Self-Acting' Mule: A Study of Skill and the Sexual Division of Labor," *Social History* 11, no. 3 (Oct. 1986), 319–43; William Lazonick, "Industrial relations and technical change: the case of the self-acting mule," *Cambridge Journal of Economics* 3, no. 3 (Sept. 1979): 231–62.

[44] Mike Williams, with D. A. Farnie, *Cotton Mills in Greater Manchester* (Lancaster: Carnegie Publishing Ltd., 1992), 9; Christine MacLeod, *Heroes of Invention: Technology, Liberalism, and British Identity, 1750–1914* (Cambridge and New York: Cambridge University Press, 2007), 161.

became known, was sometimes called the "Iron Man." At first, the mules that employed it could spin only coarse yarns. Finer counts broke more often and benefited from more attention. The older hand-operated mules became known as "mule jennies," perhaps inadvertently memorializing the Hargreaves-style jenny that had always been a nonpowered machine. The mule jennies were still needed to spin the finer yarns – anything better than 60 count, even as late as 1860. But the alternative had been made operational, and stood ready as a weapon in capital's fight against labor. In November 1836, in Preston, 650 spinners struck, leaving 30 mills and their mules standing idle. Manufacturers brought in Iron Men to replace the flesh ones, replacing labor costs with capital investments. Only 367 operatives were needed to mind the retrofitted mules, and of course these were new hires, trained only in the operation of the new machines. By early spring, the entire mule spinners' trade union at Preston had disintegrated. In Glasgow, early in 1837, thousands struck, and were easily replaced with Iron Men.[45]

The result of the new device was devastating. Mule spinning remained the work of skilled men – the shape in which the spun yarn gathered on the spindle remained a matter of regular machine adjustment, and this classified the work as skilled masculine labor. But automating the heaviest moment made possible larger mules with more spindles – from 200 to over 1300 spindles per mule by the end of the century. Even these larger machines required fewer workers, and by the 1840s, mule spinners' wages were half what they had been when they revolted in 1826. The new machines cost much more to run – they used 60 percent more energy than Crompton's mule had done in order to spin a comparable quantity of cotton. Less labor, more coal: adopting the new machine meant shifts in the inputs, in the costs and factors of production. The relative values of economic factors changed as machinery was adapted into a new production system.[46]

The Iron Man presents a case in which workers and their collective actions, their class identity, shaped the technological system and the development of machines themselves. Steam and the self-acting mule together contributed to making workers interchangeable, just as other applications of the engine had freed factories from waterfalls. The first generation of industrialists, who had already made significant investments in waterpower, continued to reap its advantages: the Gregs installed self-acting mules at Quarry Bank but powered them with the

---

[45] Harold Catling, *The Spinning Mule* (Newton Abbot: David & Charles, 1970), 48–51, 62–67; Malm, *Fossil Capital*, 66–68, 152–53.
[46] Freifeld, "Technological Change," 324–28; Farnie, *Cotton Mills*, 9; Malm, *Fossil Capital*, 65–68, 152–53.

waterwheels. The family regularly tested the relative efficiency of the two power sources: waterfalls still ran cheaper than coal in 1849.[47] There were still choices available in the industrial system. The range of factors influencing industrialization – the types of fibers, the kinds of machines, the source of power, the sex and age of the worker and his or her home and family situation – appear as a shifting array of possible arrangements within a single, recognizable structure – a matrix of accumulation. One choice set others into motion, and early investments in waterpower paid dividends for decades. As industrialization accelerated, the association of one technical choice with another became more solid. Although the Iron Man self-acting mule could be worked with water, it was more often linked to steam engines, as they spread industrial production to new locations and sectors.[48]

Thermodynamics

The steam engine also gave rise to a new branch of science, a fresh understanding of nature: the concept of energy. As steam engines became more important to economic growth in the nineteenth century, their workings became the basis of scientific investigation and a new worldview based on energy exchange. Thermodynamics developed from watching steam engines work – not the reverse. Technology was not the result of scientific knowledge. Instead scientific theories were based on the workings of machines. Sadi Carnot became known as the father of thermodynamics for his 1824 reflections on the operations of an ideal heat engine. Before, heat had been considered a chemical element, as oxygen, hydrogen, and uranium are – it was called "caloric." Usually found in fluid form, when caloric was mixed with other chemicals, it participated in the chemical reaction. Associated with the French eighteenth-century chemist Antoine Lavoisier, caloric disappeared in the nineteenth century when it was replaced by thermodynamics, which views heat not as a chemical element, but as an expression of energy. Since the Industrial Revolution, thermodynamics has, in turn, mathematized the concept of energy and made it functional, something that can be measured and put to work in the world.[49]

[47] Malm, *Fossil Capital*, 65–68, 88–89.   [48] Scranton, *Proprietary Capitalism*.
[49] Donald S. Cardwell, *From Watts to Clausius: The Rise of Thermodynamics in the Early Industrial Age* (Ithaca, NY: Cornell University Press, 1971); Crosbie Smith, Ian Higginson, and Phillip Wolstenholme, "'Avoiding Equally Extravagance and Parsimony': The Moral Economy of the Ocean Steamship," *Technology and Culture* 44, no. 3 (July 2003): 443–69; M. Norton Wise and Crosbie Smith, "Measurement, Work, and Industry in Lord Kelvin's Britain," *Historical Studies in the Physical and Biological Sciences* 17, no. 3 (1986): 147–173; Sadi Carnot, "Reflections on the Motive Power of

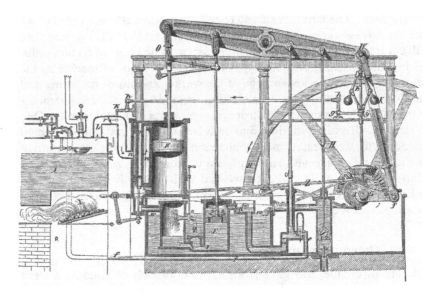

Figure 4.3 James Watt's steam engine utilized a separate condenser to keep the steam warm when the engine cooled and the resulting vacuum set the beam in motion. Patented in 1769, this machine achieved industrial might in the nineteenth century. Observing its operations inspired thermodynamics and the new science of energy. Courtesy of ZU_09/DigitalVision Vectors/Getty Images.

Technological change preceded and inspired scientific investigation and discovery, and new ways of thinking about the physical world developed from the process of industrialization. As the meaning of the word efficiency shifted from a type of cause to a measure of outputs from inputs, so did the concept of energy emerge to compare outputs across power sources. The concept of energy as something measurable, held in comparable ways in waterfalls and coal, provided the crucial conceptual basis of arguments that involved comparisons of efficiency across systems.[50] It was this sort of analysis that laid the basis for those men who were beginning to claim, for example, that a weaver using John Kay's flying shuttle had required the output of five to eight spinners, more than his traditional family, and thus viewed the machine as causing

Fire and on Machines Fitted to Develop That Power," pamphlet (Paris: Chez Bachelier Libraire, 1824).
[50] Jennifer Karns Alexander, *The Mantra of Efficiency: From Waterwheel to Social Control* (Baltimore: Johns Hopkins University Press, 2008), 15–32.

proto-industrialization. This understanding of energy viewed work as one thing, divisible by the hour and comparable to coal. The new concept of efficiency came in handy as the new machinery shifted from product to cost competition, and scientific understandings of the physical world developed in ways that fit the new production systems.

Industrialized Mobility

Steam engines ran more than textile mills. At about the same time that steam came to dominate Britain's industrial economy, the mechanism was adapted to transport goods and people among networks of places. Canals had played an important role in the early phases of industrialization, and now steam engines became a systematic part of a new mode of transportation: the railroads that used steam engines to move goods on rails. Railroads had first been tried to carry coal from the mine (at Darlington, in fact, just north of Yorkshire, where John Marshall had gone seeking flax machinery) to its market. In the early days, people piled up on top of the coal to catch a ride. But the system of railroad technology eventually coalesced into more familiar forms. The Liverpool and Manchester Railway opened in September 1830 and prohibited horse-drawn vehicles on its iron rails. Unlike the canal system, passage on the railroads was restricted to carriages supplied by the enterprise itself, by the railroad company. Steam-powered locomotives would be the vehicles used to haul raw materials, passengers, and finished goods, between the two Lancashire towns. The project had been promoted by businessmen at both ends of the line, including McConnell's partner John Kennedy. Advocates for the railroad had argued that the canals had a monopoly in linking the two ends of the Lancashire cotton industry, and that growth would benefit from competition in the form of the new technology. Steam borrowed from the existing structures of trade and transportation, and the revolutionary railways resulted. When railroads applied steam engines to the movement of goods, industrialization was clearly spreading beyond textiles – even beyond manufacturing.[51]

The steam engines, the spinning mules and Iron Men, the railroads and systematic understanding of heat as energy: all this new technology multiplied the effects of conventional work. So too did the financial cycle spread the effects of credit, and one firm's overexpansion could ripple across the economy and thrust everyone into the results of a downturn. As

---

[51] Philip John Greer Ransom, *The Victorian Railway and How it Evolved* (London: Heinemann, 1990); Mokyr, *Enlightened Economy*, 214–16; D. S. Barrie, "The Liverpool & Manchester Railway: A Centenary of World-Wide Interest," *Railway and Locomotive Historical Society Bulletin* 22 (May 1930): 46–49.

inchmeal mechanization in textiles had larger effects, older institutions creaked into new forms. Machinery's origins in product differentiation were becoming cost competition, and fossil fuels served capital's needs. As "the diffusion of machinery came to pose a sustained challenge to established labor hierarchies, provoking industrial unrest," established labor structures gave way to wages for minding machines.[52] Workers and their efforts to protect themselves against exploitation or redundancy were not merely episodes of resistance to mechanization. They are part of the story of how technology changed, and in the nineteenth century, of how industrialization matured into mass production.

As industrial capitalism solidified, workers continued to seek representation in Parliament. The long-simmering reform movement had begun to boil. In 1831, petitions urging Parliamentary reform circulated in the Midlands and the North, with tens of thousands of signatures on each one. A new Whig majority in the House of Commons responded by proposing a Reform Act that would provide seats to represent the new industrial towns like Manchester, with their swelling populations. The Act attempted to excise rotten boroughs, whose representatives were usually controlled by wealthy landlords. And it enlarged the franchise to more male heads of household, with provisions that gave the vote to about 20 percent of the male population. The Reform Act passed the House in September 1832 by more than 100 votes – but it failed in the House of Lords, who struck down the reform. Riots ensued. A more moderate bill passed in 1832 that provided Parliamentary representation to the new northern towns and extended voting rights to middle class men.[53] In 1833, too, Parliament passed an act regulating the use of child workers, their ages, and the hours of their toil. The importance of children in factories and the proportion of them in the industrial workforce was already dropping as the system matured and wages prevailed.[54]

Reverberations

Where did the goods go, the cotton products of industrial mills? From the start, the cotton industry of Lancashire, protected in its domestic markets, had been competing with Indian textiles around

---

[52] Griffiths, Hunt, and O'Brien, "Inventive Activity," 893.

[53] Boyd Hilton, *A Mad, Bad, and Dangerous People?* (Oxford: Oxford University Press, 2006), 420–24; Rose, *Gregs of Quarry Bank*, 123.

[54] Clark Nardinelli, "Child Labor and the Factory Acts," *Journal of Economic History* 40, no. 4 (Dec. 1980), 741–42, 745; Katrina Honeyman, *Child Workers in England, 1780–1820: Parish Apprentices and the Making of the Early Industrial Labour Force* (Aldershot, England and Burlington, VT: Ashgate Publishing, 2007).

the world – especially in the Atlantic trade. In America, British-made fabric was already outpacing Indian imports in the 1780s, as industrialization took shape. Lancashire cottons also did well on the west coast of Africa, where the English had long sold cloth to buy slaves. It was there, on the African coast, in the toils of the slave trade, that cotton cloth from Manchester overtook Indian goods after 1806. The tides of global commerce were shifting, and England's long apprenticeship in the business of cotton cloth was finally resulting in mass production that was beginning to supersede England's old competitor.[55]

The next step was to sell British cotton products directly into India. This happened only slowly and in intermittent steps. Some capitalists tried to industrialize Indian production: a vertical mill for spinning cotton yarn and weaving it into cloth was in operation in the 1830s, and it ran 100 power looms. By the 1840s, the mill had given up weaving – the five functioning steam engines had by then been shifted to power only the spinning operations. But it still spun yarn in counts from 20s to 50s, and the lower quality were cheaper than British imports while the better were the same price.[56] Product differentiation was giving way to global price competition.

British industrialization did not destroy the Indian textile industry immediately. In the nineteenth century, however, European mass production hit the Indian system "like a jack-hammer ... sudden, forceful, and unrelenting." In 1813, Bengal alone still exported 4.75 times as much value in textiles as Britain did. A few years later, India's importation of foreign textiles had increased, though weavers in India still supplied most of the cloth to its vast low-cost domestic market. The first phases of British industrialization aided their enterprise. Cheap, plentiful, regularized yarn, spun by steam-powered Iron Men in Manchester, bolstered Indian industry by keeping Bengali weavers competitive with British

---

[55] Joseph E. Inikori, *Africans and the Industrial Revolution in England: A Study in International Trade and Economic Development* (Cambridge and New York: Cambridge University Press, 2002), 428, 447–48, table 9.9; H. V. Bowen, *The Business of Empire: The East India Company and Imperial Britain, 1756–1833* (Cambridge: Cambridge University Press, 2006), 242–45.

[56] P. J. Cain and A. G Hopkins, *British Imperialism: 1688–2015*, 3rd ed. (Abingdon and New York: Routledge, 2016), 95, 303; Prasannan Parthasarathi and Ian Wendt, "Decline in Three Keys: Indian Cotton Manufacturing from the Late Eighteenth Century," in Giorgio Riello and Prasannan Parthasarathi, eds. *The Spinning World: A Global History of Cotton Textiles, 1200–1850* (Oxford and New York: Pasold Research Fund and Oxford University Press, 2009), 399; Parthasarathi, *Why Europe Grew Rich*, 227–28; William Bolts, *Considerations on India Affairs, Particularly Respecting the Present State of Bengal and its Dependencies*, 2nd ed. (London: Brotherton and Sewell, 1772), 194–95.

products. Through the 1850s, the pattern persisted, where factory yarns from Britain were woven in India to sell for the subcontinent's domestic consumption. But designs, dyes, patterns, and markets proved malleable as industrialization in the north of England redivided the processes performed in different parts of the world. In 1817, the Bengali imports of finished fabric began to increase, to 141,000 from 34,000 pieces just a year earlier. The invasion accelerated after 1839. In the first half of the nineteenth century alone, Indian cotton production declined by about a third. A major uprising against the East India Company in 1857 led to direct rule by the British crown in 1858 and the start of the formal British Raj, the British empire in India, under Queen Victoria.[57]

The nineteenth-century transition to steam power kicked industrialization into a higher gear. Yet even this specific case of technological change carries all the complex contingency and multi-directional causation of the larger story. Steam power provided by the fossil fuel, coal, became one of the defining characteristics of the industrial age and the capitalist system partly because a few thousand mule spinners in Lancashire wanted higher wages. Bigger capital investments and higher operating costs replaced truculent labor. The adoption and use of steam engines in textile mechanization broke the limitations of location and scale and suited mass production and price competition. Industrially made British textiles had begun to encroach on the markets of their old rival, the cotton textile industry of India, which had inspired their industrialization in the first place. Industrialization was never a linear process, and the causal relationships between capital and labor, between technological change and its local and global context, were complex. Causation moved in more than one direction as the industrial system reached toward completion.

[57] Parthasarathi and Wendt, "Decline in Three Keys," 405; Indrajit Ray, *Bengal Industries and the British Industrial Revolution (1757–1857)* (London and New York: Routledge, 2011), 65, 68–71, 75, 84, table 3.11; Douglas M. Peers, *India under Colonial Rule: 1700–1885* (London and New York: Routledge, 2013), 23–27.

Suggested Readings

Cardwell, Donald S. *From Watts to Clausius: The Rise of Thermodynamics in the Early Industrial Age*. Ithaca, NY: Cornell University Press, 1971.

Cookson, Gillian. *The Age of Machinery: Engineering the Industrial Revolution, 1770–1850*. Woodbridge: Boydell Press, 2018.

Griffin, Emma. *Liberty's Dawn: A People's History of the Industrial Revolution*. New Haven: Yale University Press, 2013.

Jeremy, David J. *Transatlantic Industrial Revolution: The Diffusion of Textile Technologies between Britain and America, 1790–1830*. Cambridge, MA: MIT Press, 1981.

Malm, Andreas. *Fossil Capital: The Rise of Steam Power and the Roots of Global Warming*. London and New York: Verso, 2016.

Mohanty, Gail Fowler. *Labors and Laborers of the Loom: Mechanization and Handloom Weavers, 1780–1840*. New York and London: Routledge, 2006.

Ray, Indrajit. *Bengal Industries and the British Industrial Revolution (1757–1857)*. London and New York: Routledge, 2011.

Smail, John. *Merchants, Markets, and Manufacture: The English Wool Textile Industry in the Eighteenth Century*. New York: St. Martin's Press, 1999.

Thompson, E. P. *The Making of the English Working Class*. London: Victor Gollancz, 1964.

Tunzelmann, G. N. von. *Steam Power and British Industrialization to 1860*. Oxford: Clarendon Press, 1978.

5   The Vertical Mill

Victorian Britons loved Paisley shawls. Originally woven in Kashmir – in the intricate pattern taken from Persia, from the fleece of a mountain goat imported from Tibet – these warm woolen luxuries had been the gifts of royals, coveted in Russia and Persia and across the Ottoman Empire since the 1500s. In the seventeenth century, European textile printers imitated the pattern, but the shawls were tricky to use. Women who had travelled to India knew how to wear them, while men who worked for the East India Company sent them home as gifts. (See Figure 5.1.) But the Calico Acts kept them rare in eighteenth-century Britain, where, just as in cotton, the prohibitions against imports inspired imitators and protected innovators. By 1777, merchants from Edinburgh to Norwich were making imitations of the prestige items. But Paisley, just outside Glasgow, reached market supremacy early in the nineteenth century.[1]

Paisley is the name of a Scottish village. That tells the whole story of the Industrial Revolution so far, and going forward. European fabric producers competed with an Asian specialty textile – first as a luxury item, later with goods made in mass quantities – until the substitute obliterated the industry it was devised to imitate. In the case of Paisley, the alternative even became the name of the design. Paisley village is also a real place and part of this history. Its textile business had experienced and participated in the accumulation of capital and the elaboration of domestic production in the early modern period. Its products then helped reorganize world trade as Britain developed a formal empire in India. Paisley was also a site where the story reaches its end as industrialization reiterated itself and settled into patterns: the move from domestic to factory production (of weaving, this time, rather

---

[1] Chitralekha Zutshi, "'Designed for Eternity': Kashmiri Shawls, Empire, and Cultures of Production and Consumption in Mid-Victorian Britain," *Journal of British Studies* 48, no. 2 (Apr. 2009), 421–25; Marika Sardar, "Indian Textiles: Trade and Production," Oct. 2003, *Heilbrunn Timeline of Art History* (New York: Metropolitan Museum of Art, 2000–); Jonathan Eacott, *Selling Empire: India in the Making of Britain and America, 1600–1830* (Williamsburg, VA, and Chapel Hill: Omohundro Institute of Early American History and Culture and University of North Carolina Press, 2016), 285–86.

Figure 5.1 Illustrated Victorian magazines taught people how to wear unfamiliar clothes and new fashions, such as Paisley shawls. Courtesy of ZU_09/iStock/Getty Images.

than spinning); the shift from product innovation to price competition in the process of mechanization; the development of industrial class structures and identities; and the role of consumer culture and preferences in shaping production systems. In Paisley, too, the newer phases of mechanization can once again be seen as part of those longer trends that separated consumption from production and regularized the flow of raw materials and finished goods worldwide.

Mechanizing Weaving

Paisley villagers had been weaving fine cloth pieces for centuries. The place had a flourishing local linen industry between the Glorious Revolution of 1660 and the 1707 act that joined England, Scotland, and Wales into one United Kingdom. By 1710, the town was known for producing cloths called "Bengals," a striped mixture of cotton and linen, presumably to compete with Indian imports. Paisley's eighteenth-century textile business illustrates the preindustrial capitalism that flourished before mechanization. Domestic weavers who had accumulated some savings, or even packmen (peddlers and distributors), sometimes set up workshops to manufacture and market Paisley products. The town specialized in silk after 1759 – one man owned six hundred looms for weaving silk gauze. By 1784, making silk gauze employed 5,000 weavers in Paisley and its surrounds. By 1790, silk had given way to fine cotton muslins, the same type that Samuel Oldknow was making in Stockport in that decade. Linked into luxury cloth networks, eighteenth-century Paisley weavers were already imitating popular and expensive imports. By 1803, this product line included shawls woven with Asian designs. Still a textile town, in 1806, some 88 percent of the town's manufacturing output was cloth.[2]

Paisley's history up to this point reminds us that weaving was still done by hand, decades after the fabled mechanization of spinning, and so it would continue for another generation. Powerlooms did exist. Francis Cabot Lowell powered looms with waterwheels in America in 1815. In Britain, there were 2,400 powerlooms operational in 1813, about 14,150 in 1820, and fewer than 60,000 in 1829. By 1833, however, there were 100,000 powerlooms at work, their rapid adoption taking place in the same decade that Iron Men were replacing labor in the spinning sector,

---

[2] John Parkhill, *The History of Paisley* (Paisley: Robert Stewart, 1857), 21, 31–32; A. Dickson and W. Speirs, "Changes in Class Structure in Paisley, 1750–1845," *Scottish Historical Review* 59, no. 167 Part 1 (Apr. 1980), 55–56; Tony Dickson and Tony Clarke, "Social Concern and Social Control in Nineteenth Century Scotland: Paisley 1841– 1843," *Scottish Historical Review* 65, no. 179 Part 1 (Apr. 1986), 53–54; Eacott, *Selling Empire*, 285–88.

and boom went bust, and steam became the prime mover of the British economy. It was in the 1820s that handloom weaving peaked in Britain from its rapid expansion during the industrialization of spinning: there were 240,000 handloom weavers in Britain in 1829, more than four times the number of powerlooms. Not only had the number of British cotton handloom weavers tripled between 1795 and 1833, they also still out-numbered the men and women employed in spinning altogether. Thereafter, the decline of handloom numbers was quick, plummeting from the peak in 1829 to 55,000 weavers in the early 1850s. Their ranks shrank to around 10,000 in the early 1870s, and they finally disappeared as an occupation in the 1880s – a century after the mechanization of spinning had industrialized and begun the mass production of yarns into counts.[3]

The genealogy of powerloom development does not resolve itself as easily into a myth of progress, hung around a series of successive machines, as does the mechanization of spinning. Scholars have traced the patents back to 1678, and, of course, the mythological version of the Industrial Revolution began with John Kay's flying shuttle of 1733 that mechanized one aspect of weaving and thereby receives credit for creating a bottleneck that initiated a cascade of spinning innovations some dec-ades later. Generally, the family tree is rooted in 1785, when the Reverend Edmund Cartwright patented a power-driven loom. It was heavy and slow; later improvements, by Cartwright and others, resulted in various devices of varying success. Some have identified a powerloom that worked in Stockport, associated with William Radcliffe, "soon after 1800," while others say the powerloom was "perfected in around 1820," although Francis Cabot Lowell had his system operational in 1815. Richard Roberts, the Manchester engineer who had devised the headstock that automated mule spinning, had also patented a powerloom in 1822 that "combined the best features of previous patents into a cast iron frame."[4] Additional micro-inventions made powerlooms work bet-ter, faster, or cheaper in the ensuing decades. One stopped the loom automatically if a shuttle needed refilling or a yarn broke and rang a bell

[3] Geoffrey Timmins, *The Last Shift: The Decline of Handloom Weaving in Nineteenth-Century Lancashire* (Manchester and New York: Manchester University Press, 1993), 35, 185; Andreas Malm, *Fossil Capital: The Rise of Steam Power and the Roots of Global Warming* (London and New York: Verso, 2016), 69–70; E. P. Thompson, *The Making of the English Working Class* (London: Victor Gollancz, 1964), 192.

[4] Gillian Cookson, *The Age of Machinery: Engineering the Industrial Revolution, 1770–1850* (Woodbridge: Boydell Press, 2018), 19; Joel Mokyr, *The Enlightened Economy: An Economic History of Britain 1700–1850* (New Haven: Yale University Press, 2012), 96; Patrick O'Brien, "The Micro Foundations of Macro Invention: The Case of the Reverend Edmund Cartwright," *Textile History* 28, no. 2 (1997), 216.

to get the attention of the operative. This allowed each worker to mind more looms at once. Of course, improvements in preparing processes, especially beaming and dressing the warp, also helped make new looms run.[5]

The mechanization of weaving was a diffuse and complicated process partly because weaving was not one single task but rather was highly differentiated by type of fiber, finished product, intended pattern, and other factors. At the same time, historical contingency also shaped technical choices and the rate at which different industries employed power-looms. The worsted merchants of Yorkshire had been heavily capitalized from the start, which gave them surplus funds to invest in new machinery: They were buying powerlooms in the 1820s, well before the smaller-scale woolen makers shifted into factory production. Worsted merchants had a lot of money lodged in raw materials, and disliked the inefficiencies of putting-out work; wool merchants generally ran smaller concerns and found the flexibility offered by domestic producers profitable. In cotton, the rate of powerloom adoption remains a subject of debate. The sheer size of the cotton industry – its volume of raw materials and finished goods – means it had more looms than the fleece-based industries. But some scholars say that despite the innovations in mass production associated with cotton spinning, most cotton firms stayed small through the mid-nineteenth century. Even a profitable small business might lack the resources to invest in the newest machinery, and, of course, the putting-out system had grown into networks of weavers available to pay by the piece.[6]

This chapter offers a rough sketch of the ways that handloom weaving disappeared and powerlooms spread across British textile production. As looms powered by inanimate sources were made to work for various types of weaving, the men who employed at-home weavers were either operating on too small a scale for mechanization, or they were making small batches of fine goods that needed more weaving skill. The persistence of handloom weaving and the eventual adoption of powerlooms were also entangled in the politics of Victorian Britain, and all these different threads can be found woven together into the history of nineteenth-century Paisley. Before returning to the Scottish village and its handloom politics, however, first consider the technical and economic reasons that

---

[5] Cookson, *Age of Machinery*, 229; Timmins, *Last Shift*, 149; Mokyr, *Enlightened Economy*, 82–113.

[6] Cookson, *Age of Machinery*, 19–20, 291; Maxine Berg, *The Age of Manufactures: Industry, Innovation, and Work in Britain, 1700–1820* (Oxford: Basil Blackwell, in association with Fontana, 1985), 209–17; Timmins, *Last Shift*, 155, 185.

handloom weaving survived through the first half-century of textile industrialization.

Handloom Innovation

Technical considerations are part of the reason that handloom weaving remained a dynamic industry for so long. In the first three decades of the nineteenth century, powerlooms wove only lesser-quality cloth made of coarse yarn. Better fabrics came from hand-powered looms. Fine yarns broke more easily and therefore needed the weaver's attention more often to fix. Spun wool was frequently more fragile than the more heavily processed worsted yarns, which went more easily into the powerloom process. In the finest cottons, where yarn breakage was more common, handloom weaving persisted, as it did in making some blends and certainly in the weaving of silk, through at least the 1830s. Even among handloom weavers, significant differences in fibers, looms, skills, and the quality of their products counted. These product differences mattered as much as did cost considerations even as the technology changed. Handloom weavers, or the men who had built the putting-out merchant firms that paid them, responded to the increasing availability of powerlooms by switching to weaving finer goods – especially silk – or more convoluted patterns in more colors.[7]

Remember that Samuel Oldknow, an Arkwright partner, had more than 450 weavers in his employ by 1786. Thirty years later, in 1816, McConnel & Kennedy supplied yarns and received woven pieces from more than a thousand men. Other Manchester manufacturers put out yarn to over 1,200 additional weavers in northeast Lancashire alone, and the satellite villages of Cottonopolis (including Oldham and Bolton) also experienced population explosions from the 1770s on.[8] Towns had their own hinterlands: Stockport mills employed 1,000 weavers outside the town, while in 1816 a single spinning company in Preston employed "a whole countryside of handloom weavers," which included most of the firm's 7,000 workers. Most merchants probably paid only a few hundred domestic workers, and some Bolton manufacturers had just ten or twenty. Not all of these weavers lived in the country. In early nineteenth-century

---

[7] Duncan Bythell, *The Handloom Weavers: A Study in the English Cotton Industry During the Industrial Revolution* (Cambridge: Cambridge University Press, 1969), 28; Timmins, *Last Shift*, 102, 148, 185–86.

[8] George Unwin, *Samuel Oldknow and the Arkwrights: The Industrial Revolution at Stockport and Marple* (New York: Augustus M. Kelley, 1968), 45–46; Bythell, *Handloom Weavers*, 29–30; Malcolm I. Thomis, *The Town Labourer and the Industrial Revolution* (London: B. T. Batsford, 1974), 5n1.

Manchester, for example, 145 cottages just off Oldham Road provided space for more than 600 handlooms, and similar weaving colonies appeared in other Lancashire towns. There is some evidence that urban handloom weavers made fancier goods than their rural counterparts.[9]

There were economic reasons as well for the expansion of handloom weaving during industrialization: It was not merely persistence. Putting-out merchants were starting up or expanding weaving workshops, powered only by muscle and apprentices, as they had done in the guild era. One of these was the father of Andrew Carnegie (the famous American steel industrialist of the late nineteenth century), who had more looms than sons, and emigrated with his family from Scotland to America when his enterprise went under. Failure was not unique to handloom weavers starting workshops; remember that four out of every five of the manufacturing firms of Manchester also failed between 1780 and 1815. Conversely, at least a few nineteenth-century hand-powered workshops made their masters rich men in their communities. A hybrid business structure that combined domestic weaving and factory spinning was actually quite common, and the economic benefits seem clear. Weavers who worked in their homes carried more of the risk of production. Those who picked up yarn from the mill and carried their finished cloth back to the merchant reduced the firm's transaction costs. Domestic work fit into industrial production systems. In 1841, two-thirds of Britain's cotton operatives worked in factories, and half the woolens and worsted workers did. The rest worked at home, or in nonpowered workshops.[10]

The old difficulties of the putting-out system had not disappeared. Embezzlement and quality control remained serious issues, and centralizing weaving with steam power was intended to solve these problems. Observers at the time viewed the powerloom as a means to control labor rather than increase efficiency or decrease costs: in a period of turbulent unrest, domestic weavers stood outside the discipline of the factory system, and this was threatening to industrialists with long experience with machine-breaking and other customary expressions of grievance. As one witness described, "The chief advantage of power looms is the facility of executing a quantity of work under more immediate control and management, and the prevention of embezzlement, and not in the reduced cost of

---

[9] Bythell, *Handloom Weavers*, 28–30, quotation at 30; Timmins, *Last Shift*, 42, 63–64, 73–78.

[10] "Carnegie Started as a Bobbin Boy," obituary, *New York Times*, Aug. 12, 1919; Boyd Hilton, *A Mad, Bad, and Dangerous People?: England 1783–1846* (Oxford: Clarendon Press, 2006), 23; Timmins, *Last Shift*, 154–55, 172–73; Mokyr, *Enlightened Economy*, 96–97.

production."[11] Just as in the mechanization of spinning, economies of scale were more a result of powerloom weaving than its intended aim.

But handloom weaving in the nineteenth century resembled the mythological version of domestic production mostly in the location of work and the source of power. Work at home preserved some features of the domestic and putting-out systems of production, but other features were disappearing. Although many handloom weavers continued to work in their homes, their status as independent producers suffered. Most were new recruits to the weaving industry rather than proud patriarchs, inheritors of a domestic tradition. They took up weaving because there were few jobs in the factories for grown men. Women and children went out to work in spinning mills or coal mines, and their earnings contributed to the household economy, rather than supplying raw materials to the loom. In some instances, women and children even wove at home, contributing household production to the family's subsistence, but reducing any vestigial status of manly independence still associated with the job.[12] Most handloom weavers were part of the emerging working class, as were the factory operatives who minded machines in the mills.

Paisley's long history with the production of luxury goods makes it a good example of the transformation. At first, Paisley weavers used sophisticated harness looms to weave the sinuous twisted teardrop in silk, cotton, or a blend of the two. In 1812, they adopted the ten-box lay, which got them nearer the quality of Kashmiri weaving. Their investors never stopped adapting: various dobby and Jacquard looms were employed over the decades to weave the eponymous patterns, and these were not adapted to use with inanimate power until the 1860s ("and even then took time to perfect").[13] In 1833, French weavers of ornate patterns began to adopt Jacquard looms, and Paisley weavers were using the new machine in 1850. This nineteenth-century French loom used a chain of cards, each with holes punched in it for hooks to pass through, to lift and lower successive patterns of warp yarns. The Jacquard loom's punched cards functioned as a kind of a programmable heddle and could be used to weave complicated patterns: on a Jacquard loom, a weaver could even produce full-size, nonrepeating pictures and mass produce tapestry. Up

[11] Quoted in Maxine Berg, *The Machinery Question and the Making of Political Economy, 1815–1848*, rev. ed. (Cambridge and New York: Cambridge University Press, 1980), 242.

[12] Berg, *Machinery Question*, 250; Timmins, *Last Shift*, 108, 119–34.

[13] Timmins, *Last Shift*, quotation at 150; Dickson and Clarke, "Social Concern and Social Control," 53–54; Thomas W. Leavitt, "Fashion, Commerce and Technology in the Nineteenth Century: The Shawl Trade," *Textile History* 3, no. 1 (1972), 53, 57.

to the 1860s, additions to these looms allowed weaving in more and more colors – and they usually ran without any inanimate power.[14]

Working-Class Politics

Despite the skill involved in weaving such elaborate designs, Paisley weavers worked on looms supplied by other people, and this made them working class by definition. Paisley was also a village with a history of radicalism, and its weavers participated in the transition from traditional remedies to political solutions. In 1812, the handloom weavers in Paisley went on strike. They wanted all the area manufacturers to agree to pay equivalent prices for finished goods. They failed to win the desired outcome as cost competition battled with quality improvements. The defeated weavers organized into trade unions in response, even while unions were still illegal. The decades after 1815 saw these weavers' unions regularly fighting – armed conflict in the streets – against the authorities. Paisley weavers responded to the August 1819 Peterloo massacre in Manchester with five days of rioting and the violence reached its climax in 1820 – finally put down by military and militia forces. No wonder Iron Men seemed like a reasonable response to worker violence among the Manchester manufacturers, with their capital invested in mills and machinery. In 1824, when the Combination Acts were repealed, trade unions made legal grew even more bold. Industrial action still shaded into violence, which darkened the reputation of the unions. But working-class actions were shifting their focus from industrial to political activities.[15]

Chartism – a political movement aimed at voting rights for the working class – took its name from the People's Charter of 1838 whose main provisions included universal male suffrage, a secret ballot, annual elections, the elimination of property qualifications for government service, and pay for Parliament Members. Yet Chartists in the northern industrial cities did not shrink from force. They carried weapons when they paraded, demonstrating their willingness to fight for political rights. Their taste for older ways appears not only in their methods but also in their goals and ideologies. Their Charter showed the patriarchal nature of working-class rhetoric and ideology. Powerloom weaving was fast becoming the domain of female machine minders. Resistance to the factory system of production therefore relied on mythologized visions of the alternatives, and depicted the handloom weaver as the natural father.

---

[14] Janet Delve, "Jacques Vaucanson: 'Mechanic of Genius'." and Delve, "Joseph Marie Jacquard: Inventor of the Jacquard Loom;" both in *IEEE Annals of the History of Computing* (Oct.-Dec. 2007), 94–97 and 98–102.

[15] Dickson and Speirs, "Changes in Class Structure in Paisley," 66–69.

Domestic production systems had glorified the male head of household, and his family was his workforce. The independence embodied in domestic textile technologies became the ideal for which the radicals reached. The "increasingly gendered world of organized labour," according to Chartism's historian, conceived of men as breadwinners and women as wives and mothers rather than factory workers. Paisley was one of the places where Chartism held its greatest appeal, and some of the movement's greatest leaders were powerful handloom weavers in the village.[16] The older system of production was a resource for the organization of industrial workers.

In Paisley, the local history of the Chartist movement also shows that radical handloom weavers and local industrial capitalists were not yet entirely separate categories. The radical working class hated the "political economy" associated with Adam Smith, which advocated the division of labor and mass production. But on a local level, laborers and capitalists were linked both socially and economically. There were many Dissenters from the Church of England among both groups, and religious preferences and values brought them together. In addition, Paisley capitalists who put out work into people's homes in the nineteenth century were also making finer products on ever-evolving looms. Small-batch production runs made fashionable goods responsive to market, and kept many firms small in order to be nimble. Most people who invested capital in handlooms still relied on independent domestic workers to operate the machinery. They were industrial capitalists but also putting-out merchants, and their efforts went to product innovations rather than mass production. These local contingencies show another example how the industrial system drew on preindustrial structures, even as it transformed them.[17]

The basis of the preindustrial system was gone, however. Families no longer much made their own yarns, and they wove for others, for pay, not to make cloth to wear or use as a bedcovering. Instead they earned money to buy clothes. Reports from the time suggest that most households in the north of England had stopped making their own fabric for clothing, especially adult clothes, by the 1830s. Homespun seemed shabby, not artisanal. Cloth woven by other people, whether in their own home on a handloom, or by powerlooms in factories, was readily available for those who earned pay in the factories or domestic service. "Higher incomes

---

[16] Chase, *Chartism*, 5, 12, 52–62, quotation at 43; Carol E. Morgan, "Women, Work and Consciousness in the Mid-Nineteenth-Century English Cotton Industry," *Social History* 17, no. 1 (Jan. 1992), 29–31.

[17] Berg, *Machinery Question*, 234; Timmins, *Last Shift*, 36; Dickson and Speirs, "Changes in Class Structure in Paisley," 59–64.

were more likely to be spent on increased quantities of shop-bought fabrics than on increased quantities of raw flax or wool" to be worked up at home.[18] The separation of consumption from production was largely complete, even as the mechanization of production was still taking place.

High-fashion small-scale production systems were as unstable as the demands they satisfied, however. Paisley town suffered during the volatile nineteenth-century boom-and-bust cycles of emerging industrial capitalism. Depressions in the textile trade struck the town during the French Wars and afterwards, with recurring troughs in 1793, between 1811 and 1812 and again between 1816 and 1822. The bust of Britain's financial system in 1825 meant another depression in 1826–1827, and again in 1829, and 1831, and 1837. Competition for jobs that paid ever-lower wages drove earnings down further, and domestic weavers' rates kept dropping. By 1834, even the best handloom weavers earned less than a third of the pay in 1791. By that same year, in 1834, Paisley weavers were producing £1 million worth of shawls every year. Paisley shawls had become the standard gift for brides. People wore the shawls in different color combinations for the different seasons. The Paisley shawl was popular across class divisions in nineteenth century Britain, an Asian design industrialized.[19]

Echoes and Interactions

At the same time that Paisley weavers were producing Asian-styled shawls in their homes in Scotland, India's weavers and merchants were also improving their products. Kashmiri shawls still dominated the international luxury market. They used the very best silks and fibers and fleeces, woven in increasingly elaborate designs. Three hundred shades of yarn were not unheard of. Posh Kashmiri goods inspired competitors, an international nineteenth-century shawl business within which Paisley operated. French, British, and Kashmiri merchants bought and sold both raw materials and finished shawls, and borrowed and adapted each other's designs to meet consumer trends. The finest shawls from Kashmir could fetch over £200 in the period from 1820 to 1840, and Paisley firms sent agents to London to copy the newest designs, as Oldknow had gone decades earlier. By the 1830s, Paisley could have duplicates on the market

---

[18] Styles, *Dress of the People*, 145–46, quotation at 146; Carolyn Steedman, *Labours Lost: Domestic Service and the Making of Modern England* (Cambridge and New York: Cambridge University Press, 2009).

[19] Dickson and Speirs, "Changes in Class Structure in Paisley," 55, 67–69; Leavitt, "Fashion, Commerce and Technology," 53.

in eight days, fast fashion for a tenth the price. At first, Paisley shawls had warps made of worsted or imported silk; after 1830, firms in Bradford in Yorkshire began machine spinning cheaper yarns that lowered the price of inputs at Paisley. In the same decade, Paisley manufacturers also began supplying their weavers with lower quality weft yarns. This maneuver not only dropped the input cost of raw materials, it also required weavers to weave more to achieve the same pay.[20]

Handloom weavers, still paid by the piece, served well the industry's transition from the product innovations that imitated imports from India to cost competition that lowered workers' pay. Between 1825 and 1850, Indian producers competed with the Paisley industry on this basis, making cheaper goods with more standardized designs and fewer colors. Competing for the bottom of the market did not help their reputation. After 1850, Scottish manufacturers were adopting the Jacquard loom from France, and weaving in as many as fifteen colors. This provided a reasonable facsimile of the original luxury goods, produced on hand-powered looms, from cheaper machine-made yarns. By that time, it was obvious to consumers that Kashmiri shawls were no longer equal to Paisley, and the Scottish shawls were readily available, with a range of goods that everyone could buy and wear. Copying the luxury helped make a larger market for cheaper goods. By 1870, the export trade in Kashmir shawls from India to Europe had declined to almost nothing.[21] Mass production was the result, though it had not been the goal of new machinery. Instead, product and process innovations shaped one another: more colors and looms designed for the making of more luxurious goods resulted in a range of products that were also cheaper and more readily available. Technology intended to improve quality – to standardize the traits of raw materials and vary the characteristics and output of finished goods – resulted in cost competition and mass production. Economic historians say this happened in Lancashire cotton in the 1790s, while threads of it can still be spied in Paisley shawls in the 1830s.[22]

The contest between Paisley and Kashmir reminds us that the process by which product innovation became cost competition had an imperial context and a mercantilist justification. Paisley's triumph over Kashmir's

[20] Zutshi, "Designed for Eternity," 424; Leavitt, "Fashion, Commerce and Technology," 53–57.

[21] Zutshi, "Designed for Eternity," 424; Leavitt, "Fashion, Commerce and Technology," 53–54, 57, 59–61.

[22] Trevor Griffiths, Philip A. Hunt, and Patrick K. O'Brien, "Inventive Activity in the British Textile Industry, 1700–1800," *Journal of Economic History* 52 (Dec. 1992), 893; Cookson, *Age of Machinery*, 27–28.

shawls was a small part of the larger pattern, in which the British Raj of 1858 relied on knowledge of the territory acquired by the efforts of earlier institutions. The East India Company and a firm of British merchants in Calcutta sent an emissary into Kashmir in 1822, to try to gain access to the networks that supplied the special Tibetan wool. The Lancashire vet and horse breeder they sent did not succeed in sourcing the raw materials, but he "methodically" studied the Kashmiri manufacturing processes, from cleaning the wool to how to wash the finished product. His 1822 trip to Kashmir is yet one more example of how Company efforts to participate in trade gave way to learning Asian production methods, to replicating them to secure advantage among European rivals.[23] Participating in trade became replacing exotic products with domestic substitutes, and making better products became making more of them, more cheaply. Paisley in the 1830s had become another example of the way that the old Indian Ocean trade and its institutions supplied the business and product knowledge and the economic incentives that inspired industrialization.

The Jewel in the Crown

Victoria was crowned Queen of England in 1837. It was a tough economic year, the end of a boom cycle, and the Bank of England raised interest rates, which sparked economic panics around the world. She was 18 years old, and she would rule the United Kingdom until she died in 1901, well after the end of our story. She was a real human being and also a mythological character, a monarch, the Queen of England – among other titles. She ruled the Raj and became Empress of India in 1877. Her name emblazons a period of British history. Five years into her reign, in 1842, during yet another economic slump, the young Queen Victoria publicly wore a Paisley shawl. She also bought cloth woven with the pattern, to make some dresses. She sparked the craze for Paisley to a mass level, and received credit for a revived economy the next year. Royal fashion had inspired consumer fascination with calicoes in the seventeenth century, when Charles II wore his waistcoat sewn from Indian fabric. It still played a role in the nineteenth century, as industrial production matured into recognizable form.[24]

It was in the 1830s, too, that British politics began to acknowledge the social class structures of the industrial towns, organized around the capital investments of Northern textile manufacturers. Parliamentary

---

[23] Zutshi, "Designed for Eternity," 427–28.
[24] Dickson and Clarke, "Social Concern and Social Control," 50; Hilton, *Mad, Bad, and Dangerous,* 499; Parthasarathi, *Why Europe Grew Rich,* 31.

Reform in 1832 had expanded suffrage but did little to remove the landed gentry from dominating the nation's political institutions. The land-owners who controlled Parliament had spent the decades investing in their estates, not in factories. Bankers in London were "disinclined to involve themselves in manufacturing," while merchants put their capital in shipping firms and other services, rather than mills and machinery. Manufacturers occupied a distinct social position as a result: not a Liverpool Lord, whose fortune came from commerce, but a Manchester Man. As working-class identity grew, capital also recog-nized its distinct interests and flexed its political voice. As the laboring classes marshaled their efforts from customary violence toward trade unionism, punctuated by Peterloo, so too did capitalists transform the institutions that had brought them into being. As Britain's industrial manufacturers took steps toward political power, their social identity cohered around a shared set of myths describing their origins in terms of machinery and business daring.[25]

Inventing Invention

Edward Baines, Jr. (1800–1890), Leeds newspaperman, religious Dissenter, and son of a man of the same name and profession, was a prolific author. In 1835, he published *A History of the Cotton Manufacture in Great Britain*. The title set the terms: cotton manufacture in Great Britain, not wool or steam power. Baines liked to attribute technological change to individual inventions and inventors, but he also identified the causes of mechanization within the history of world trade and the response of Britain's established textile sector to the competition presented by Indian cloth. He wrote at the moment that British manu-facturing was overtaking markets from Indian textiles, and recalled for his readers the threat that Indian cottons had posed to "our ancient woollen manufacture." In his view, the inferiority of India's cotton cultivation and manufacturing methods made "exaggerated and absurd" the terrors of the British wool industry that had received "almost superstitious venera-tion" from the government.[26]

Perhaps he was projecting his 1835 comparison onto the past, when Indian textiles had a world-wide reputation, and represented serious

---

[25] P. J. Cain and A. G. Hopkins, *British Imperialism: 1688–2015*, 3rd ed. (Abingdon and New York: Routledge, 2016), 52, 83; Isabella [Mrs. George Linnaeus] Banks, *The Manchester Man* (London: Hurst and Blackett, 1876); "The 'Manchester Man' and the 'Liverpool Gentleman,'" *Age*, 13 May 1868 (second edition), p. 7.

[26] Edward Baines, Jr., *History of the Cotton Manufacture in Great Britain* (London: H. Fisher, R. Fisher, and P. Jackson, 1835), 75–89, quotations at 77, 79.

competition to the British industry. But Baines disapproved of the protection that had been extended over the established textile business of the eighteenth century, within which the cotton industry had developed. His generation preferred laissez-faire, and his history reflected his worldview. He dissociated industrialization from the mercantilism that had supported its success. Even though his text acknowledged the contextual causes of machine adoption, Baines still treated machines as causing change. By emphasizing individual devices and their genealogies, Baines encompassed a history of global trade within the names of a few British men.

Baines was not the only one – pamphlets and books had already begun celebrating heroic inventions, and would continue to do so into the future. The format of these accounts mythologized industrialization. An early example was William Radcliffe, who tried a powerloom at Stockport, and published his results in 1828 as the "Origin of the New System of Manufacture, Commonly Called 'Power-Loom Weaving.'" It was this book that promoted the idea that John Kay's flying shuttle caused an eighteenth-century bottleneck in domestic production. Radcliffe even supplied the almost exact number of five-to-eight spinners required to supply the new looms of the 1730s – however piecemeal their adoption had been.[27] It was the Northern industrialists themselves who celebrated their genius, who taught the Victorians to identify and praise inventions and ascribe them to individuals, who thus received credit for the recent growth of British industry and its accomplishments. Later observers noted that the greatest invention of the nineteenth century was the invention of how to invent – by which they meant to indicate an endless improvement of machinery. But it was in the nineteenth century, too, that the idea of inventions as the causes of change began to take hold.[28] But, of course, the processes of invention are not so simple, and include not only the sorts of tweaks and modifications involved in making a machine work, but also those continuities and shifts in human behavior and social organization that account for any machine's successful operation.

The invention narrative of the 1830s also described real events, the effects of which were becoming clear. Mass production served Britain's

---

[27] William Radcliffe, *Origin of the New System of Manufacture, Commonly Called "Power-Loom Weaving"* (Stockport: James Lomax, Advertiser-Office, 1828), 60.

[28] Christine MacLeod, *Heroes of Invention: Technology, Liberalism, and British Identity, 1750–1914* (Cambridge and New York: Cambridge University Press, 2007), 8; MacLeod, *Inventing the Industrial Revolution: The English Patent System, 1660–1800* (Cambridge and New York: Cambridge University Press, 1988); Mokyr, *Enlightened Economy*, 81.

industrial and imperial dominance, as more people in more parts of the world became markets for things made by others, and steam power made mass production possible across a widening swath of the economy. Steam-engine horsepower increased by probably 50 percent in Britain in the two years between 1835 and 1837 alone. Growing railway networks account for some of this expansion: the Liverpool and Manchester Railway had signaled the start of a new cycle of investment. Opened in 1830 to carry cotton from America to the mills, and finished cotton goods back to the port, the venture pointed toward even more economic growth. By Victoria's coronation in 1837, the railway mania had already begun. Some railway companies even eventually purchased about a quarter of the existing canals. Railroads bought the canals for right-of-way and turned them into roadbeds for the tracks, so parts of the new transportation system lay right on top of the existing one. The boom in railway building inspired more than a thousand schemes proposed by 1845. When Parliament approved new lines, financial speculation followed. The British railway system of the twenty-first century reflects these origins, with networks and links dating back at least to the nineteenth century – and sometimes even earlier, going back to the canals.[29]

Steam power was everywhere after 1830, signaled by the railways, spreading across sectors. Increasing numbers of engines certainly aided the adoption of powerloom technology across the nineteenth century. Power-woven cloth was not always cheaper than that made by hand, but its output was more predictable and so were its quality and characteristics. Taken together with the new capacity supplied by steam engines – capacity wasted if unused – the self-acting spinning mule and the power-loom together brought more of the production system into factories where it could be linked up and driven as one unit. The two machines together, and their power sources, made the factory system spread even further. By 1838, four-fifths of English cotton mills were likely powered by steam engines. People who invested in the new machinery found it had to keep running to recoup its capital cost. With the downturn that accompanied Victoria's coronation in 1837, manufacturers kept powerlooms running and let handloom weavers go. Because powerlooms could usually be run by women, the work appealed less to men accustomed to higher status. Men unused to factory discipline were also hard to manage, and expensive. After the Irish potato harvest failed in 1845, Irish immigrants came across the water, desperate for work in English factories. They were easily slotted into the new weaving systems. These newcomers were

[29] Timmins, *Last Shift*, 94–96; Philip John Greer Ransom, *The Victorian Railway and How it Evolved* (London: Heinemann, 1990), 11, 253–55.

almost universally reviled in England, and their presence further diminished the job's reputation.[30]

Back in the 1820s, in the early days of powerloom adoption, the Manchester manufacturers who added power weaving to their existing spinning operations needed space to accommodate the new machines. The weight of the looms and their vibrations required single-story construction, a large shed typically with a serrated or saw-toothed roof in which the large vertical surfaces were made of glass, to maximize the light that fell onto the work. In some cases, the ridges fell to troughs that carried rainwater into gutters and from there to the steam engine, usually lodged in the basement of the mill. It was hard to find flat sites in the crowded city for the new weaving sheds. Purpose-built mills that combined spinning and weaving processes began to appear in the early 1830s. These were the famous vertical mills that integrated all the technical processes of textile production. Raw materials went in and finished cloth came out. Production from preparation through spinning to weaving was powered from a single inanimate power source, by belts that were turned from the shafts rotated by the steam engine. The noise of their operation was deafening, and each weaver managed four looms at one time. It was a long way from John Kay's flying shuttle, patented a hundred years earlier. By 1841, 58 percent of all workers in Lancashire cotton factories worked in mills that combined spinning and weaving.[31]

Integrated production was not unique to the cotton sector during the swift ascent of steam power in the 1830s. Other fibers experienced similar consolidation of tasks under one roof. John Marshall, the linen manufacturer of Leeds who had hired the tinsmith-engineer Matthew Murray who worked out machinery for heckling flax in the 1780s and 90s, also participated in the nineteenth-century blossoming of industrialization. In the early years, he had sold some of his yarn to independent producers but also maintained a putting-out business for weaving. In 1793 he increased the number of his handlooms from 43 to 81, and by 1798 he had 150 weavers at work in nearby towns. These people bought their looms on installment from him, and together used about half of his mills' output of yarn. Then, the French Revolution and Napoleonic Wars made his fortune. In 1836, Marshall broke ground for the construction of a vertical mill of grandiose construction. Modeled on the ancient Egyptian Temple

---

[30] Malm, *Fossil Capital*, 75; Berg, *Machinery Question*, 23; Berg, *Age of Manufactures*, 252–53, Timmins, *Last Shift*, 92–97, 161–67; Morgan, "Women, Work and Consciousness," 24.

[31] Mike Williams, with D. A. Farnie, *Cotton Mills in Greater Manchester* (Lancaster: Carnegie Publishing Ltd., 1992), 76, 84–85; Malm, *Fossil Capital*, 76. Queen Street Mill, in Harle Syke, Burnley, Lancashire (visited by the author in summer 2013).

of Edfu, the flax mill of Leeds was fronted with heavy stone columns in the shape of flax plants. Inside the two-acre main room, the interior columns that supported the skylit roof, also cast in the form of flax, were hollow pipes that carried rainwater from the roof down to the steam-engine (see Figure 5.2). In linen as in cotton, in worsted and eventually wool, machines were available and deployed across the textile production processes. Yet machines were only one element of the production systems that spanned agricultural production and retail distribution.[32]

Some vertical mills even included housing for their workers. This was not a new strategy: Samuel Oldknow had built up Marple, near his Stockport spinning mill, in the eighteenth century; Robert Owen's spinning schemes had included a benevolent village in New Lanark, Scotland; and Frances Cabot Lowell had adopted the system in America, to house unmarried women from the farms of New England. Such paternalistic structures within a factory complex show how the patriarchal household of domestic production could be applied to the new business structures of industrialization. Titus Salt, for example – a wool merchant of Yorkshire – in 1851 built a new stone town just outside Bradford. His complex contained not only a six-story spinning mill and a weaving shed that covered two acres, but also terraces of stone houses intended for the healthful accommodation of the mill's workers. He modestly named his village Saltaire, after himself. His steam engines at Saltaire spun wool sheared from alpaca sheep that grazed in South America. The exotic fiber required considerable adaptation of machinery before the vertical production system could be made to work. This was the newest version of industrialization: in the middle of the nineteenth century, with steam power and vertically integrated, completely mechanized factories, using a fiber produced very far away, housing the workers in a domestic system led by a patriarch manufacturer, now supported by coal-burning steam engines. In 1869, Queen Victoria made Titus Salt a baronet for his accomplishments.[33]

Vertical mills that combined spinning and weaving on one site, driven by one power source, required large capital investments. They epitomize industrial capitalism. They also hearken back to the household. When mechanized power-driven spinning was new,

[32] W. G. Rimmer, *Marshalls of Leeds: Flax-Spinners, 1788–1886* (Cambridge: Cambridge University Press, 1960), 35, 53, 69, 203–07.

[33] James Burnley, *Sir Titus Salt, and George Moore* (London: Cassell & Co., 1885).

INTERIOR OF MARSHALL'S FLAX-MILL.

Figure 5.2 John Marshall's 1837–41 flax mill in Leeds utilized the latest innovations. Weaving sheds like this one were flat single-storied rooms with overhead windows to provide light, and they were adopted in cotton production at about the same time as Marshall built this one to weave linen out of flax spun on site. The exterior (photographed by the author in 2016) was modeled on the ancient Egyptian Temple of Edfu, with its columns in the shape of flax plants. The interior columns, likewise modeled on flax plants, were hollow pipes that carried rainwater down to the steam engine that ran the machines through a system of shafts and gears, as in Figure 2.4. bauhaus1000/ DigitalVision Vectors/Getty Images.

different towns around Manchester had specialized in different products, as Bolton was a fine-spinning town and Oldham spun coarser counts. This had reflected a proto-industrial system that divided textile processes into different houses, spinning from weaving, a division of labor at the level of the village rather than the household. Now, as weaving sheds were added to spinning mills, the household that did all the textile tasks reconstituted itself on a larger scale, in the factory. And the factory – composed of multiple buildings, a complex of production sites that divided labor but still kept all the tasks within one mill complex – itself echoed the distant plantations of the American South, where the cotton was grown to enter that faraway factory. The new industrial production system drew on technological and physical assemblies that already existed, even as some of those systems were destroyed. As weaving mechanized and domestic production disappeared, it emerged in new forms or transformed into new structures. Family labor systems were a resource on which industrialists drew, even as they faded.

Institutional Changes

Francis Cabot Lowell used the American corporate form to finance his weaving sheds after 1815. In Britain, similar legal accommodations for capital investments can be dated to the 1824 repeal of the Combination Acts – the same repeal that had legalized trade unions. As working-class identity was solidifying into institutions of protest and politics, capital also combined into new institutions, including corporations. After Combination was permitted in 1824, Parliament received a flood of requests from companies seeking charters of incorporation. These were mostly joint-stock companies, whose stock of ownership shares could already be bought and sold. The East India Company had used the joint-stock form of business organization; in the nineteenth century, critics of the joint-stock form focused on the monopolistic privileges granted to chartered companies like the EIC.[34] Critiques were older than that: both guilds and joint-stock companies had fallen into disrepute by the late eighteenth century. Adam Smith, political economist at the University of Edinburgh, had skewered the Company and the mercantilist system in his 1776 *Wealth of Nations,* which had also celebrated early divisions of labor. His work provided the ideological basis of

---

[34] Ron Harris, *Industrializing English Law: Entrepreneurship and Business Organization, 1720–1844* (Cambridge and New York: Cambridge University Press, 2010), 199–206, 254.

laissez-faire capitalism, at the moment that manufacturers were joining their forces into political activism.[35]

So advocates of the corporate form renewed the charges against the old joint-stock companies and claimed that the earlier ways of organizing investments impeded competition, as the organizers of the Liverpool and Manchester Railway had claimed about the canals. At the same time, though, the joint-stock form served as the legal and institutional precedent that carried out their goals. The historian of industrializing English law describes the East India Company as providing a "path dependency" that structured the way corporate law developed. A series of Parliamentary Acts in 1824, 1826, and 1833 ended centuries of corporate monopolies in banking and marine insurance, while the EIC lost most of its trading privileges between 1813 and 1833. By the end of the 1830s, "the monopolistic feature so strongly linked" to these early versions of "the business corporation finally disappeared," and the path was cleared for the competitive version, organized for profit.[36] The new business form stood on the old as its advocates both diminished the contribution of the existing structure, and wrecked its legal standing.

Then, in 1844, the passage of a Joint-Stock Companies Registration Act made incorporation more accessible. Instead of requiring a special charter to create a corporation, the new law permitted joint-stock firms to act as corporations in place of their individual stockholders. Another act in 1855 provided that the liability of individuals who owned stock in a company be limited to the loss of that investment; if a firm caused an accident, for example, the victims could sue the company but not receive as recompense the private property of its individual owners. This limited liability proved a durable feature of the modern business corporation. In 1856, the Joint Stock Companies Act linked these two laws and made the modern corporation a legal structure for protecting investments and allowing them to pay returns or reveal their risks over time, outside the individual investor's activities. Over the next 14 months, more than 1,600 firms registered on that basis – more than twice the number of corporations already in existence. Incorporation had begun as a legal status granted only by special dispensation, a royal charter to town governments and the guilds that coordinated town business. In the nineteenth century,

[35] Adam Smith, *An Inquiry into the Nature and Causes of the Wealth of Nations* (London: W. Strahan and T. Cadell, 1776), especially book iv.
[36] Harris, *Industrializing English Law*, 215, 290.

it became a business structure, a legal and political institution that served industrial firms and protected their capital investments.[37]

Like so much else in our story, the nineteenth-century rise of the corporation was more the result of industrialization than its cause. In the booming cotton sector of the late eighteenth century, merchant partnerships and family firms predominated. Manufacturers plowed merchant capital back into their businesses without usually resorting to joint-stock business forms. Networks of families and friends supported industrialization – even country banks remained unincorporated, due to restrictions posed by the Bank of England. Nonetheless the joint-stock company was used widely enough in transportation projects and in shipping firms, including "overseas flows," of materials, that it is usually counted as at least some aid to industrialization. But the corporation was not a cause of industrialization, except at the remove of the charters granted to guilds, towns, and joint-stock companies like those that built canals and engaged in overseas trade. The corporate form was a result of new business types, built on investments of capital that paid out over time.[38]

More institutions to organize capital likewise took shape in the nineteenth century, including private organizations that provided rules and structures to the new business forms around industrialization. The Liverpool Cotton Brokers' Association, for example, formalized its members' long-established commerce in cotton with membership rules and transaction procedures, starting in 1841. Liverpool remained the port and the central location for the arrival of cotton bales from the southern United States, which were then sent by railway to Manchester. The eighteenth-century docks continued to serve the Atlantic trade, and American independence still relied on the export of agricultural commodities from slave-worked plantations to the old imperial mother. The cotton from the American South was manufactured in Britain in increasing quantities, and Manchester built a new exchange in 1809 to handle the business in yarn and cloth. Thus Liverpool's trade in cotton fiber linked it to the established commercial world while Manchester had become the marketing center for the whole cotton manufacturing industry. Manchester and Liverpool businessmen organized into local Chambers of Commerce during the political turmoil of the French Wars. In the 1820s and 30s, Chambers like these spread to other sites across Britain, offering a political voice that united local commercial and

---

[37] Harris, *Industrializing English Law*, 282–91; Mokyr, *Enlightened Economy*, 357–58.
[38] Harris, *Industrializing English Law*, 174, 198, 212.

manufacturing interests. Capitalists were beginning to share a group identity, and they did so by uniting for a specific political goal.[39]

Free Trade

By the 1830s, the Corn Laws stood for the whole established link between land and government, locked in place since at least the medieval period. At the end of the Napoleonic Wars, Parliament imposed these tariffs that kept the price of grain high to protect those who produced it at home – landowners whose income depended on agriculture. The Manchester Chamber of Commerce had been protesting these tariffs throughout the 1820s. The national Anti-Corn Law movement sprang from a Manchester base – there were 225 clubs in England organized in protest of the Corn Laws, most of them in Lancashire, London, and the West Riding of Yorkshire. Robert Hyde Greg, the son of Samuel Greg who had founded the Quarry Bank Mill at Styal, was one of its prominent leaders. Parliament was responsible for the high cost of bread, declared the Anti-Corn Law leagues that were spreading at the end of the 1830s. The tariffs had expressed a mercantilist political economy, in which nation-state and business worked for mutual advantage and a favorable balance of trade against other nations. The Anti-Corn Law agitators were among those groups who drew on Adam Smith's *Wealth of Nations* to marshal arguments against what he called the "mercantile system." They described their goal as "free trade," by which they meant dismantling the mercantilist protections for the landed gentry. This was a cause that also appealed to working-class unionists. Despite their opposition to the "political economy" that animated Free Trade ideology, they wanted lower food prices. The network of Anti-Corn Law agitators provides yet another example of labor and capital working in opposition, but also in tandem, in the history of the Industrial Revolution.[40]

[39] Thomas Ellison, *The Cotton Trade of Great Britain. Including a History of the Liverpool Cotton Market and of the Liverpool Cotton Brokers' Association* (1968; orig. pub. London: Effingham Wilson, Royal Exchange, 1886), 181; Cookson, *Age of Machinery*, 186–87; Robert J. Bennett, *Local Business Voice: The History of Chambers of Commerce in Britain, Ireland and Revolutionary America 1760–2011* (London and New York: Oxford University Press, 2011).

[40] Hilton, *Mad, Bad, and Dangerous*, 502–13; Mary B. Rose, *The Gregs of Quarry Bank: The Rise and Decline of a Family Firm, 1750–1914* (Cambridge and London: Cambridge University Press, 1986), 84, 128–33; Harris, *Industrializing English Law*, 206.

The Conservative party triumphed in the election of 1841 on its customary platform to protect and conserve the Church and landed gentry. Its Prime Minister, Robert Peel (1788–1850), came from a textile manufacturing family of Lancashire – his father had been named a baronet after making a fortune in the late eighteenth century by printing sprigs of parsley on calico fabric. According to the legend, Parsley Peel had worked with James Hargreaves on carding technology and later also employed his spinning jenny – he was one of the men who fought Arkwright's patents in court. The son became Prime Minister of an administration that in the 1840s encouraged foreign trade and lowered taxes and tariffs on consumer items, offsetting the revenue losses with an income tax. Peel's government thus also modernized the Parliamentary regulations on English business organization, and oversaw the legal development of the modern corporation.[41]

The economic modernization ascribed to Peel's party not only freed up capital through corporations, it also took steps to govern the emerging technologies. New regulatory frameworks were established in both the Railway Act of 1844 and the Factory Act of 1847. The Factory Act, both the way it was passed and its effects, illustrates the complicated contingency of the relationship between social and technological change. The Act limited the work stint of women and children below eighteen years of age to ten hours. Its proponents in the Ten Hours' Movement had used the logic of steam as they chose their goals: since steam engines dictated factory time, limiting the hours of even some classes of workers would (they hoped) inspire manufacturers to run factories only while the full workforce toiled. Thus the steam engine served labor's strategy. At the same time, though, limiting operations to ten hours made the waterwheels' inanimate constancy lose relative efficiency to steam engines that could be turned off. And steam power was already serving as a tool of capital in replacing labor with machinery.[42] So the technology and its regulation shaped one another as the system matured into its recognizable nineteenth-century form.

The voters expected the Conservative Party to protect agrarian and Anglican interests, but the liberal wing of the party had other ideas. Reducing tariffs on incoming consumer goods and establishing an income tax were just the start. There were factions among the Tories, and ultraconservatives fought the liberal legislators rising under Peel's

---

[41] Alfred P. Wadsworth and Julia De Lacy Mann, *The Cotton Trade and Industrial Lancashire, 1600–1780* (Manchester: Manchester University Press, 1931), 477, 482, 498n2; Harris, *Industrializing English Law*, 278–86.

[42] Harris, *Industrializing English Law*, 278–86; Griffiths, Hunt, and O'Brien, "Inventive Activity," 889; Malm, *Fossil Capital*, 186–93.

premiership. Their squabbling created an opportunity, while public opinion in the North flared against the protections. Near the start of November 1845, Prime Minister Peel suggested the suspension of the Corn Laws, either on principle, or perhaps to show he flirted with intellectual fashion. His party revolted at the suggestion. They forced his resignation but no new government was willing to take up the issue; when Peel returned to the office he brought the proposal to the House of Commons, where it was debated for more than a month. When the vote passed in 1846, the Conservative Party split into shards – the repeal passed only with the help of the opposition. It also provided a triumph for the North, for the new classes formed by textile industrialization. In repealing the Corn Laws, the industrial classes of the North had constructed an ideal around liberal capitalism, free trade, and antiprotectionism. "Free Trade" was always more mythological than descriptive. Manchester's manufacturers did not envision a world in which their colonies functioned as independent economic actors, for example, as free traders in world markets. The Crown would ensure the dependency of India, not as competitor but as colony. Manchester men also opposed the export of textile machinery to other countries. They called free market capitalism the policies and economic theories that supported their own economic interests, and imagined that these conditions that enabled their expansion had actually started their rise.[43]

Manchester men built a Free Trade Hall, a public assembly and concert venue, to celebrate their victory. Free Trade Hall rose on a location of mythical power to working-class radicals and activists: the site of Peterloo, where the political activism of labor had met the response of its government in mounted police, swinging cutlasses, and trampling crowds. To an earlier generation, the empty field had been the site where working people learned they were on their own. Now it became a structure that supported a mythological history of the Industrial Revolution. Completed in 1856, Free Trade Hall was decorated with carvings that celebrated trade and industry, commerce and a mythical history of free trade. Mills and chimneys, bags of wool: the benefits of free trade were carved in stone. The Hall commemorated how an industry nourished in the arms of government protection had now fought itself free of the embrace, while maintaining the markets and products of empire. The repeal of the Corn Laws and the celebration of that repeal in Manchester's Free Trade Hall demonstrated a new order for Victorian

---

[43] Hilton, *Mad, Bad, and Dangerous*, 502–13; Douglas A. Irwin, "Political Economy and Peel's Repeal of the Corn Laws," *Economics and Politics* 1, no. 1 (Spring 1989): 41–59; W. J. Ashworth, *The Industrial Revolution: The State, Knowledge and Global Trade* (London: Bloomsbury, 2017).

Figure 5.3 This stone medallion from Manchester's Free Trade Hall
shows a mill and chimney, and a ship representing commerce, behind
a woman seated on a bag of wool and holding the sprig of a plant,
possibly cotton, or indigo, or tobacco. Courtesy of Bob Speel.

England (see Figure 5.3). As industrialization ripened, its leaders found
political power. The old landed gentry that had ruled England since the
imagined days of King Arthur now had new rivals.[44]

---

[44] Historic England, "Free Trade Hall," List entry no. 1246666.

Manchester's industrialists assembled new myths to serve these political ends. They invented invention as the explanation of industrialization, of the remarkable transformation of the Lancashire cotton industry since the 1760s. Victorians raised statues of inventors across the land. They celebrated steam and railroads more often than spinning cotton. In Manchester especially, they "submerged Arkwright's claims on their gratitude beneath a show of admiration for Watt." Schoolchildren learned the names of inventors, and learned to ascribe to them the social and global changes that had occurred with industrialization. The mythological Industrial Revolution would rely on this genealogy. The first author to use the term linked James Watt to Adam Smith: one "distilled the doctrine of free trade; the other's contemporaneous invention had facilitated that doctrine's worldwide implementation." In this simple version, two men created the Industrial Revolution. James Watt with steam and Adam Smith with free trade had together "destroyed the old world and built a new one."[45] Ascribing change to machines, and to an imaginary laissez-faire for events that had actually grown up cradled by protective policies, was not really a lie. It simply mistook effects for causes, as Baines had done, in his assessment of Indian textiles. This confusion saw technological change as the result of invention, a flash of genius that changed the world. It rested on the belief that technology had transformed the world around it: transformations in the household and the family, shifts in the commercial systems that wove the world together. But these elements were causes of changing technology, at the same time that they were effects.

Industrialization even fulfilled the old dreams of mercantilism, while its heralds glorified free trade. After all, the material, physical reality of the Industrial Revolution involved processing at home raw materials acquired from abroad, adding value to goods bought where they were common and sold around the world. Meanwhile, just as the household shaped the factory system that then changed home into new forms, so had commerce and industry split into separate sectors. The Liverpool Cotton Brokers' Association controlled its members and their transactions, as guilds had done. The Royal Exchange building in Manchester was built by its subscribers as a place to do their business, a mirror of the Cloth Halls of merchant capitalism, transformed into an institution of industrialization.

---

[45] Arnold Toynbee, *Lectures on the Industrial Revolution* (1844), 14, 189, quoted in Christine MacLeod, *Heroes of Invention: Technology, Liberalism, and British Identity, 1750–1914* (Cambridge and New York: Cambridge University Press, 2007), 116–44, 193–95.

The Royal Exchange kept expanding with the cotton trade, until a new building was erected in the last third of the nineteenth century.[46]

Labor also relied on existing institutions, not just in trade unions but in the very organization of work in the factories. Powerloom weavers and mule spinners were still paid by the piece – paid not for the hours they worked but for how much cloth they wove, or how much yarn they spun, just as when they worked at home. From the perspective of the employer, this made sense. A weaver responsible for minding several machines still managed how fast full shuttles replaced empty ones and therefore how much time the loom worked each day and, by extension, how much cloth it wove. Piecework from domestic production systems thus carried through into factories even as machines and power dictated the pace of production. Mule spinners were also paid according to how much yarn their machines produced, and they paid their own piecers – their young helpers, often their sons – out of their own wages. The spinner who minded several machines also directed the repair of broken yarns, so the team had control over maintenance. Mule spinners trained their sons to one day take their own jobs. Thus the guild system persisted, in small but structural ways, into the steam-powered machine-driven industry.[47]

Historical Materialism

Factory owners paid their workers for how much yarn or cloth they made. Workers paid their own helpers out of their own wages. Both these arrangements satisfied capitalist production calculations while nonetheless resembling and relying on preindustrial arrangements. But the owner of the machine was still the one who had decided how powerful an engine to buy. He controlled how much coal that engine received and therefore how fast it operated. We have seen capital's numerous strategies for keeping operating costs low and responsive to demand. Owners charged operatives for steam wasted during any absence. During a downturn, they could cut hours and pay, sometimes forcing a turn-out or strike, slashing the costs they paid in wages. Now, with steam engines running Iron Men and powerlooms, the mill owner controlled how fast the machine worked, but the worker still received his pay based on how much the machine made. This created a fundamental conflict over the means of production.

---

[46] Liverpool Cotton Brokers' Association Proceedings, 1842–1851, Records Office, Liverpool Central Library, Liverpool, UK; Alan Kidd and Terry Wyke, *Manchester: Making the Modern City* (Liverpool: Liverpool University Press, 2016), 75.

[47] Robert G. Hall, "Tyranny, Work and Politics: The 1818 Strike Wave in the English Cotton District," *International Review of Social History* 34, no. 3 (1989), 447, 450; Berg, *Age of Manufactures*, 186.

It served as the touchstone for Karl Marx, economist and political theorist, whose *Das Kapital* explored, in three massive nineteenth-century volumes, the history and economic relations of industrial capitalism. In 1848, he collaborated with Friedrich Engels, who four years earlier had published a study of working-class life in Manchester, *The Condition of the Working Class in England*. The two men together published "The Communist Manifesto" in London, just as a wave of revolts crested in many European cities, bringing notice and lending credence to their ideas, collectively called historical materialism.[48]

Karl Marx analyzed capitalism as a system in which wealth could be grown. Investments were risks, but they could reap profits through the exploitation of labor, which – Marx argued – added the value that accrued to investors as profits. This surplus value became returns on investment, and thus robbed workers of the ancestral returns on their labor. Capital won the surplus value generated by labor, and dictated what portion labor received by controlling the machines and the rate of work. Relying on the examples presented by textile industrialization, Marx and Engels articulated a darkly powerful history of capitalism: the colonial sources of capital, its support by the state and ruling classes, and its fundamental conflicts with labor; the role of wages and the exploitation of workers by the capitalists' quest for profits; and the development of commodities – interchangeable, mass-produced goods – and the fetish consumers made of them. Marx viewed class conflict (indeed, eventual working-class, proletarian revolution) as the inevitable result of industrial capitalism, since the economic interests of capital and labor conflicted. Marxist analysis was the opposite side of the coin of the free trade promised by Adam Smith. Smith's 1776 *Wealth of Nations* had celebrated mass production and the division of labor, while Marx in the next century saw in the same technological transformations the destruction of traditional society and the creation of a radical working class.[49]

Marx had his own context, and his ideas were the products of his time. He witnessed the maturation of industrialization, in which steam engines were deployed into productive systems and economic relations shifted to make them work profitably. As he wrote in 1847, in *The Poverty of Philosophy*, "the hand-mill gives you feudalism, the steam-mill gives you

[48] Karl Marx, *Das Kapital Kritik der politischen Ökonomie*, 3 vols. (Hamburg: Verlag von Otto Meissner, 1867–1894); Frederick [Friedrich] Engels, *The Condition of the Working Class in England in 1844, Complete and Uncensored*, translated by Florence Kelley Wischnewetzky ([Springfield, MA]: Seven Treasures Publications, 2009; orig. pub. 1844); Karl Marx and Friedrich Engels, "Manifesto of the Communist Party," pamphlet, 1848.

[49] Rius, *Marx for Beginners* (New York: Pantheon, 2003).

industrial capitalism." In fact, the causation was more complex than that, and Marx knew it. The contest around steam power and the adoption of the Iron Man self-acting mule identifies just how much industrial capitalism gave us the steam-mill, rather than the reverse. Once adopted, the steam engine acted on the world around it, from labor arrangements to raw material supplies. That world continued to reflect its origins, however, as mule spinners paid their piecers and trained them up as apprentices to the factory job where they would be paid by how much they made, just as they would have been paid by merchants for each piece woven at home. In other writings, Marx clearly understood this complexity of causation around technological change. In Book One of *Kapital,* his analysis lodged the causes of industrialization in social change rather than machinery – in the creation of wage laborers from one-time independent artisans or tenant farmers, a process that accompanied proto-industrialization. Social changes had provided the advantage to new machinery.[50]

Work and Home

The separation of work from home and the effective end of domestic manufacturing created leisure – time away from work – and activities to fill it. The first railway line opened to Blackpool in 1840, and thus established a Victorian tradition: the seaside holiday, the amusement piers, and all the trappings of industrialized, mass-produced entertainment. The railroad sold third-class tickets that catered to "the lower orders of Lancashire society," the factory workers and mill operatives who drew on a tradition of annual pilgrimages to the long straight sandy beach and vigorous breezes. Newspapers of the 1820s and 1830s depicted hundreds of holiday-makers, "cavalcades" that used "a motley array of conveyances" to get to the coast from places around Manchester. The railway replaced their customary transport on cart and horseback, their lodgings in barns and shared beds. The working classes were not the new train's only customers. Since at least the 1770s, Blackpool had served the fashionable classes as a seaside resort; the train symbolized the industrialization of leisure. Blackpool catered for the range of social classes in industrial Britain: the gentry, industrialists, and working classes all enjoyed the sea, each in their own ambits. Like a Paisley shawl, there was a holiday made for each class of consumer. By 1846, when

---

[50] Karl Marx, "The Poverty of Philosophy," chapter 2, trans. by the Institute of Marxism-Leninism (Moscow: Progress Publishers, 1955; orig. pub. 1847); Donald MacKenzie, "Marx and the Machine," *Technology and Culture* 25, no. 3 (July 1984), 480–87.

Blackpool's line linked directly into the emerging national rail network, it already served the whole social structure of the region.[51]

Seaside holidays thus demonstrate the results of mechanization, of the separation of consumption from production, and the stability of distribution and raw material supplies. So did a new social process: the Victorians learned to shop. In the absence of the old preindustrial sumptuary laws, spending would provide visual clues to class status. Fashion had been a crucial component of the forces behind industrialization, as cheaper goods and emerging consumer groups sped the adoption and replacement of goods for the adornment of homes and bodies. Consumption had initiated industrialization, a fashion for Indian textiles lit up by King Charles II in the seventeenth century and expanding to whole new classes of consumers in the eighteenth and nineteenth centuries. The medieval markets and early modern cloth halls provided models for nineteenth-century shopping arcades, which fashionable towns built to enclose their market stalls (see Figure 5.4). These arcades were covered promenades, lined with shops, each with a window to display the luxurious goods for sale inside. Some scholars have identified the first covered passageway in Britain, intended for shopping and lit by skylights, as the Royal Opera Arcade, built between 1815 and 1817 in London. But its predecessors included medieval markets and more modern Cloth Halls, trading spaces used by merchants and guilds, while its innovations included lifting the restrictions of members. Anyone with money, in the appropriate garb, could shop in the new arcades.[52]

The Victorian tendency to associate technological change with individual machines and inventors found full expression in another iteration of the chartered market towns: the 1851 Great Exhibition in London. Its famous Crystal Palace was made of iron-ribbed glass construction – similar to the roof of a weaving shed and also typical of the Victorian railway stations. Both types of shed made use of wide expanses of glass held in place by cast-iron structural frameworks, marvels of engineering, and symbols of the new industrial age. These practical structures reached artistic glory in the Crystal Palace. The exhibition was intended partly to celebrate Victorian Britain's industrial power.[53] Manchester was even

[51] John K. Walton, *Blackpool* (Edinburgh: Edinburgh University Press, 1998), 14–24.

[52] Margaret MacKeith, *The History and Conservation of Shopping Arcades* (London and New York: Mansell Publishing, 1986), 1–3, 11, 15, 18; Prasannan Parthasarathi, *Why Europe Grew Rich and Asia Did Not: Global Economic Divergence, 1600–1850* (Cambridge and New York: Cambridge University Press, 2011), 31; Giorgio Riello, *Cotton: The Fabric That Made the Modern World* (Cambridge and New York: Cambridge University Press, 2013), 130–34.

[53] See the plan labeled "Class 11," 10 Feb. 1851, Papers of the Royal Manchester Institution, Manchester City Library, Manchester, UK.

Figure 5.4 The Victoria Quarter shopping arcade, built in Leeds in the nineteenth century, echoes the Cloth Hall where merchants bought wool pieces a century earlier. Photograph by the author.

formally named a city in 1853. But Manchester had already mounted its own Exposition of British Industrial Art six years before the London extravaganza. The 1845 Manchester exposition displayed all the wonders of textile industrialization, of spindles and powerlooms, "where iron gives arms and steam supplies animation."[54] The 1851 Crystal Palace told similar stories, amplified from the national stage, the seat of empire. The Great Exhibition covered under glass the inventions and products of the industrial age.

The familiar story of the Crystal Palace reminds readers that other histories of the Industrial Revolution covered much the same ground as this book has done. Very few of the tales told here would surprise scholars of the topic. Yet the mythical version treats machines as causes of social change, where this book has attempted to put machines at the heart of the

[54] Janet Wolff and John Seed, *The Culture of Capital: Art, Power and the Nineteenth-Century Middle Class* (Manchester: Manchester University Press, 1988), 157n39.

story: at a midway point between cause and effect, as artifacts that have both reasons and results, which are often the same thing, that itself is changing. Instead of engaging in a fruitless quest for firsts – the first invention of a mechanical spinning device of one design family or another – this study instead asks and explains how these machines came to work then. The answer always involves some elements outside the machine itself. Technology is a process of doing things, using tools or machines to do it, and that process encompasses whole worlds beyond the machine. From the Indian cotton cloths to the metalworking artisans whose skills made textile machinery possible; from the tenant farmers pushed from the land, and the pauper children relegated to the apprentice system of parish charity; from the royal consumer who sparked a trend to the women splattered with acid for wearing calico cotton cloth; to the newly middle-class customer who shopped the arcades, industrialization had changed the world. In the process of technological change, industrialization drew on resources that already existed, and made the old world new again.

Suggested Readings

Berg, Maxine. *The Machinery Question and the Making of Political Economy, 1815–1848*. Rev. ed. Cambridge and New York: Cambridge University Press, 1980.

Fox, Celina. *The Arts of Industry in the Age of Enlightenment*. New Haven: Yale University Press, 2010.

Harris, Ron. *Industrializing English Law: Entrepreneurship and Business Organization, 1720–1844*. Cambridge and New York: Cambridge University Press, 2010.

MacLeod, Christine. *Heroes of Invention: Technology, Liberalism, and British Identity, 1750–1914*. Cambridge and New York: Cambridge University Press, 2007.

Mokyr, Joel. *The Enlightened Economy: An Economic History of Britain 1700–1850*. New Haven: Yale University Press, 2012.

Ransom, Philip John Greer. *The Victorian Railway and How It Evolved*. London: Heinemann, 1990.

Timmins, Geoffrey. *The Last Shift: The Decline of Handloom Weaving in Nineteenth-Century Lancashire*. Manchester and New York: Manchester University Press, 1993.

Conclusion

The Industrial Revolution provides a useful case study of the complicated relationship between technological change and the changing world around it, and how they are bound up together and with the political goals and roles of the people and institutions involved. One part of the system devised to spin yarn did not cause all the rest; machinery was interwoven with politics, only one element in a newly emerging system – emanating from the old, becoming something new. Devices and methods developed and were chosen for expedient and economic reasons. To use a kind of historical shorthand for the political history of industrialization: mercantilism validated a prohibition on the importation of cotton cloth, within which Britain's cotton industry developed. Once this cotton industry had grown a bit, Arkwright changed the legislative framework, and got the Calico Acts repealed, which made his machinery work profitably while keeping it protected by tariffs from competition with Indian imports. This origin story disappeared from the mythological version of the Industrial Revolution when the fall of the Corn Laws incorporated the Free Trade ideology of Adam Smith and his veneration of dividing work into smaller tasks. These ideas had grown from criticizing the mercantilism that had supported industrialization. In marshaling his critique, Smith gave mercantilism a coherent identity it had lacked in practice.

This sort of circular causation between technology and the world around it appears as well in the economic context, the luxurious imports against which British domestic industry measured itself and the machines it adopted and adapted resulting in mass production, itself an outcome of the increasing consumer base, and eventually destroying the old competitor. In the changing technology, too, we find histories of infrastructure – market towns and cloth halls and corn mills, packhorses and canals, guilds and town corporations, joint stock companies and the corporate business form. Why sever Richard Arkwright's water-frame or Samuel Crompton's spinning mule from any of these? Embedded as it is, one marvels that technology has become associated with progress. Yet it so manifestly has: who wants to rear sheep and spin yarn to make cloth

185

to pound in urine and devise into all their clothing? Industrialization has freed much of the world from such drudgery. Anyone with the time and skills to read this book should acknowledge the debt to the Industrial Revolution and to those who lived through it.

Within the Industrial Revolution narrative, specific stories illustrate this complexity of causation. The factory had plantation precursors and also a history of hosiers with machines to keep secret, as well as incentives to control the time of workers and the quality of their products, in addition to employing inanimate power. Many yarns of causation wove together into the idea and the artifact of a factory. McConnell and Kennedy advised their customers and berated their suppliers in trying to get each mule running. In business correspondence, they discussed changes in machinery and the buildings that housed it, and how one influenced the other. In Yorkshire's fleece-based industries, too, the technical and economic differences between worsted and woolen production, accumulating since the sixteenth century, contributed to their relative rates of mechanization in the nineteenth century. Likewise, the adoption of the Iron Man self-acting mule as a replacement for workers demanding more pay and power demonstrates the importance of contingency in directing technological change. In all these cases, economics shaped technological systems, and vice versa. The smaller stories that compose the Industrial Revolution, as well as its larger narrative, show the complicated way that human devices fit into human worlds and purposes, at every scale of analysis.

Readers of this book likely knew from the start that technology both reflects and animates the world around it. Tracing how that complexity of causation played out in actual historical events has meant recognizing a simpler version: in a case as famous as the Industrial Revolution, the shorthand has achieved a mythical, explanatory stature. But myths themselves have histories, too, bound up in the politics (in this case) of national remembrance and self-identity of Britons in the nineteenth century, of both working-class and capitalist identity. This mythological version of the Industrial Revolution is as simple as a family tree of machinery. Its bare branches include the genealogy of spinning machines and the fuzzier mechanization of weaving, the genius of steam and James Watt's achievements, and the Luddites as a categorical example of resistance. These limbs frame a myth selected once upon a time, by men with specific goals, who sought a government that represented their interests. Easily digested, their success became collective wisdom: that machines devised by ingenious heroes transformed not only textile production, but also changed the world. In the usual story, novelty met with resistance, but triumphed due to superior efficiency. The skeleton of this story ignores the interweaving of systems that defies straightforward linear analysis – the flesh of the body, which this book has taken pains to sketch.

The history of technology is not a list of inventions that changed the world. The phrase "technology in the Industrial Revolution" still too often implies that a "wave of gadgets that swept over Britain" caused the shifts we have witnessed between the first and the last chapter. Technology remains the black box to which so many explanations finally resort. But machines do not explain mechanization. Instead, inventions are the things that this history of technology has tried to explain.

In order to explain how and why textile producers and merchants incorporated and made operational a few famous machines in late eighteenth century Britain, this book has described the systems and networks within which new machines were tried, and some made operational. Always, we have looked for the reasons new machinery worked. The explanatory elements have been numerous, a universe too vast to sketch but barely here: buildings and workplaces and transport links and power sources, raw materials from the mines and the plantations, children from poorhouses and workers from farms and African coasts, financial brokerage, and Indian empire. This story is bigger than machines, which themselves require explanations, which can only be found outside any individual device. Local contingencies can be found in Manchester, which was itself embedded in worlds of trade and commerce, buildings and flagstoned squares of exchange, ports and ships and slave markets, cotton plantations and Bengalese weaving villages, Kashmir and Paisley. To pluck these strings, to suggest the intricate web of causation that reverberates among the changing textile machines and the changing world around them, has been the main task of this work. A vast new structure of labor and families, households and firms, finance and global trade, was by the 1840s creaking into existence, fashioned from the remnants of an older world it was helping destroy, as the industrialized world came into being.

Appendix: Alternative Examples

Using the three-part definition of industrialization articulated in the introduction, this book has focused on arguably the most famous case of technological change in history, the textile industrialization that happened in a few counties in England, over a period of time about as long as a human life, which has achieved a mythological status as the Industrial Revolution. Actually, similar processes were happening at about the same time in a lot of places and sectors. We have touched on others, as textile history both built on and contributed to their trajectories – new technologies of transportation and power supplies in particular, with analogous cases suggested in finance, pottery, and machine-making. Readers could have been asked to linger over the spectacular innovations occurring in Portsmouth navy yard, as military and maritime innovations contributed greatly to Britain's success in the French Wars, around which so many of our textile businessmen had to negotiate their businesses. The narrative could have focused more intensely on the canals that linked Yorkshire and Lancashire, and the productive resources of the Pennine mountains that link them, to the sea. But these systems have been cast as adjunct technologies to the changes in textile production and distribution, rather than stars of the show. A slight shift in emphasis would produce a different narrative, sometimes with different outcomes. Not every episode of industrialization resulted in mass production and distribution of consumer goods; not all technological change is successful – and sometimes, as in textile industrialization, where they ended is not where they intended to go.

Dedicated historians have produced extraordinary work documenting alternative cases. Not just cotton and not just textiles: the production of paper, pottery, iron goods, pins, buttons, and many other products were undergoing similar processes in the late eighteenth and early nineteenth centuries. Some industrialists applied steam earlier, some mass produced standardized consumer goods, some mechanized under the wing of the government or the military as a customer. Each case offers subtle differences, but each conforms to the same three-part definition of industrialization employed here. Mechanization is, of course, the face of the Industrial Revolution, the mythological version that guides our grasp. The separation of production from consumption also holds true across multiple cases, including the division of work from home, of raw material supply from processing or from using and discarding. Yet old structures sometimes reconstituted on different scales – towns did tasks once delegated to individual household members, and factories relied on the separation of spinning and weaving before they came together again under the impetus of adopting new power systems. Finally, industrialization both relied upon and stimulated the regular flow of materials between supply and demand. And each case of industrialization met this three-part description in the decades between the 1760s and the 1840s.

The hero of pottery industrialization is Josiah Wedgwood, who sold dishes by sample at his London warehouse to consumers who expected their purchases to match exactly what they had chosen. In 1765, in response to their picky needs, after long experimentation, he produced a cream glaze. The emerging middle classes loved the classical purity of the white tableware – especially after the Queen placed an order. Of course, Wedgwood the inventor (like James Hargreaves and James Watt) had a business partner, a textile merchant trained in Manchester, based in Liverpool, with connections to markets in North America and the West Indies. Wedgwood established factories and reorganized labor, managing workers with a system of bells. The firm innovated flexible batch production to produce short runs of fashionable goods. Wedgwood not only revolutionized mass production of glazed china tableware, he also remained all his life a hand potter, and extended craft processes into experimentation and product testing. To keep its goods current and fashionable, the firm trained and employed teams of designers: Handicraft remained a crucial component of Wedgwood's business. Like Richard Arkwright, Wedgwood stands on others' shoulders: Staffordshire potters had been dividing labor in new ways since the 1750s, and also using better raw materials, in response to blue-and-white Asian imports – a story sketched in Chapter 1. Eventually, pottery manufacturers adopted steam power.[1]

While most coal in eighteenth century Britain was used domestically in stoves, rather than industrially in engines, it was crucial to early industrialization in some sectors other than textiles. Ironworking in particular

---

[1] His partner was Thomas Bentley. Regina Lee Blaszczyk, *Imagining Consumers: Design and Innovation from Wedgwood to Corning* (Baltimore and London: Johns Hopkins University Press, 2000), 4–11; Robin Holt and Andrew Popp, "Josiah Wedgwood, Manufacturing and Craft," *Journal of Design History* 29, no. 2 (Jan. 2016): 99–119.

relied heavily on coal. Along the River Severn, the East Shropshire coal-field provided raw materials for turning iron into pots and pans that found ready markets in America and also traded for slaves in Africa. Shropshire ironworkers also had methods and machines for boring out iron cylinders, useful for cannons or for steam engines, which were still used mostly where coal was plentiful. The iron industry depended on the earlier canals, which expanded after the Duke of Bridgewater's 1765 canal carrying coal from his mines into Manchester sparked a canal craze, mirrored later by the mania for building and investing in railways that carried its goods and its coal. Shropshire's industrialists even had wooden railways before iron rails. Some used waterpower, though, and some built villages for workers. In 1779, Abraham Darby, a Quaker ironmaster, built the dramatic and delicate Iron Bridge across the Severn River's gorge. It advertised his ironworks and symbolized the age – it stands for industrialization as surely as does a spinning mule. When ironworking declined in the nineteenth century, the region's pottery industry remained and used the region's coal to fire its kilns. Among the Shropshire products were the typical fireplace tiles that surrounded coal fires in Victorian parlors. These homes were sites of consumption rather than production, as industrialization had established.[2]

In France, the industrialization of papermaking took familiar forms. Labor reorganization accompanied piecemeal mechanization in the eighteenth century, and portions of the work remained handicraft, artisanal processes, into the nineteenth century. National regulation of the technical processes, established in 1671, buttressed the "modes" – the customs and working habits of the journeymen. The guild structure meant that journeymen who had completed their apprenticeships tramped from mill to mill, learning and teaching. Their independence and their allegiance to their modes made them troublesome to the masters, and reducing journeymen's power was a crucial part of the industrialization process in the 1790s. Seeking employees rather than apprentices, mill masters altered the organization of their production systems to keep their workers fixed to time and work discipline; eventually they adopted machinery for some processes and thus made workers into machine-minders, though some were highly skilled and respected. Some of their mills were more like villages, with places to live and worship. The traditional division of labor into tasks done by specific age and sex groups made papermaking "family work" even when performed in the master's

[2] Barrie Trinder, *The Industrial Revolution in Shropshire* (Stroud, Gloucestershire: Phillimore & Co., 2016; orig. pub. 1973), 2, 4, 11, 27, 31, 54–59, 61, 99–105; Richard Hayman and Wendy Horton, *Ironbridge: History and Guide* (Stroud, Gloucestershire: History Press, 2009), 9, 88, 93–104.

workshop. In many cases, industrialists paid the wages for the whole family's work to the husband and father, and thus maintained patriarchal family structures.[3]

In French textile production, too, the eighteenth century saw new systems and older structures fit together in a range of possible configurations. For example, Lyon specialized in luxury silk production, and laws regarding labor organization turned the city itself into a sort of vertically integrated, highly capitalized mill. Known as the "grande fabrique," enshrined in law in 1744, this production system regulated who did what parts of the process. Women paid by the piece did distinct tasks in their homes or in workshops, and merchants coordinated the processes and marketed its products. In Rouen, on the other hand, guilds persisted into the Revolutionary age, many headed by female masters in 1775. Cloth halls regulated the trade, and the textile production systems wore the "full regalia of medieval craft ideals." Historians have spent a great deal of energy trying to answer the question why Britain industrialized and France did not. It might be more fruitful to ask which parts of the industrial system worked in specific locations, and which did not take hold, a model employed by Daryl Hafter in comparing the systems of Lyon and Rouen. Side by side, both old and new systems were employed in making French textiles for market, before the Revolution swept away so many traditional structures.[4]

Even in England, as some scholars have pointed out, there was "nothing inherently industrial about the Industrial Revolution. Since 1800 the productivity of agriculture has increased by as much as that of the rest of the economy, and without these gains in agriculture modern growth would have been impossible."[5] Crop rotation systems, the use of turnips and clover to increase the productivity of existing fields, the adoption of new plows adapted from Chinese models, even the enclosure of common lands into privately managed property – all contributed to rising agricultural productivity between the seventeenth and the late nineteenth centuries. It was this increasingly productive agriculture that permitted the demographic shifts with which this book began, the growing, better-fed British generations of the eighteenth century. Some people even had enough resources for buying things at market – for shopping – and their

---

[3] Leonard N. Rosenband, *Papermaking in Eighteenth-Century France: Management, Labor, and Revolution at the Montgolfier Mill, 1761–1805* (Baltimore and London: Johns Hopkins University Press, 2000), 31, 50–60, 66–68, 87, 93–95, 151, quotation at 93.

[4] Daryl M. Hafter, *Women at Work in Preindustrial France* (University Park: Pennsylvania State University Press, 2007), 89–99, 123–51, quotation at 100.

[5] Gregory Clark, *A Farewell to Alms: A Brief Economic History of the World* (Princeton and Oxford: Princeton University Press, 2007), 193.

desires spurred producers to meet their demands. The process did not end with industrialization: in the nineteenth century, fertilizer from Chile entered into British agricultural systems, and growth continued to accelerate.[6]

Technological change is not a process that really comes to an end. This book has ended in the 1840s, with the mechanization and the spread of mass leisure, itself a product of the gulf between work and production on one side and the growth of mass consumption on another. By then, men realized that something extraordinary was underway, and expanding. Industrial cloth production had achieved a moment of technological closure. They knew how to do it: where to get raw materials that validated their investment in machinery, what to make and how to sell and distribute it. All these industrialization stories share in common a transition to mass production aided by widening markets – more people to buy goods, in shifting configurations of world trade. In these accounts, machines occupy a middle position between cause and effect. Getting them working was another story.

---

[6] Jan de Vries, *The Industrious Revolution: Consumer Behavior and the Household Economy, 1650 to the Present* (Cambridge and New York: Cambridge University Press, 2008); G. E. Mingay, *The Agricultural Revolution: Changes in Agriculture, 1650–1880* (London: Adam & Charles Black, 1977); Mark Overton, *Agricultural Revolution in England: The Transformation of the Agrarian Economy, 1500–1850* (Cambridge: Cambridge University Press, 1996); Peter M. Jones, *Agricultural Enlightenment: Knowledge, Technology, and Nature, 1750–1840* (New York: Oxford University Press, 2016); William Edmundson, *The Nitrate King: A Biography of "Colonel" John Thomas North* (New York: Palgrave Macmillan, 2011).

Bibliography

Manuscripts

Samuel Crompton Papers, Bolton Archives and Local Studies Service, Bolton, UK.

Stanley Dumbell Papers, Special Collections, University of Liverpool, UK.

Gott Business Papers, Brotherton Library Special Collections, University of Leeds, UK.

R. Greg and Co. Papers, Manchester Central Library, Manchester, UK.

Leeds White Cloth Hall Papers, Brotherton Library Special Collections, University of Leeds, UK.

Liverpool and Manchester Railway Papers, Liverpool Records Office, Liverpool Central Library, Liverpool, UK.

Liverpool Cotton Association, Liverpool Records Office, Liverpool Central Library, Liverpool, UK.

Records of Marshalls of Leeds, Special Collections, Brotherton Library, University of Leeds, UK.

McConnell & Kennedy Papers, John Rylands Library, University of Manchester, UK.

Samuel Oldknow Papers, John Rylands Library, University of Manchester, UK.

"Plans of All the Spinning Factories within the Township of Manchester." Bound Manuscript Volume, Dated by Watermarks to 1822, Includes Plans of McConnel and Kennedy's Mill. John Rylands Library, University of Manchester, UK.

Rathbone Bros. & Co. Papers, Special Collections, University of Liverpool, UK.

Royal Manchester Institution Papers, Manchester Archives and Local Studies, Manchester Central Library, Manchester, UK.

Archives of Soho [Boulton & Watt], Library of Birmingham, UK.

Philip Sykas to Barbara Hahn, email in the possession of the author, March 9, 2016.

Wyatt Manuscripts, Library of Birmingham, UK.

Museums and Heritage Sites

Ancoats History Project, Manchester, UK.

Bradford Industrial Museum, Yorkshire, UK.

Bridport Museum, Dorset, UK.
Crewkerne and District Museum, Somerset, UK.
Dean Clough Mills, Halifax, Yorkshire, UK.
Helmshore Mills Museum, Rossendale, Lancashire, UK.
Leeds Industrial Museum at Armley Mills, Yorkshire, UK.
Manchester Museum of Science and Industry (MOSI), Lancashire, UK.
Museo della Seta, Como, Italy.
New Lanark Village, Lanarkshire, Scotland, UK.
Paisley Museum and Art Galleries, Scotland, UK.
People's History Museum, Manchester, Lancashire, UK.
Quarry Bank Mill, Styal, Lancashire, UK.
Queen Street Mill, Harle Syke, Burnley, Lancashire, UK.
Saltaire, Yorkshire, UK.
Silk Museum and Paradise Mill, Macclesfield, Cheshire, UK.
Verdant Works, Dundee, Scotland, UK.
Whitchurch Silk Mill, Hampshire, UK.
Working Class Movement Library, Salford, Lancashire, UK.

Periodicals

Hunt's Merchant's Magazine
Melbourne Age
New York Times

Printed Primary Sources

Acts of Parliament of the United Kingdom
Baines, Jr., Edward. *History of the Cotton Manufacture in Great Britain*. London: H. Fisher, R. Fisher, and P. Jackson, 1835.
Banks, Mrs. George Linnaeus [Isabella]. *The Manchester Man*. London: Hurst and Blackett, 1876.
Bolts, William. *Considerations on India Affairs, Particularly Respecting the Present State of Bengal and Its Dependencies*. 2nd ed. London: Brotherton and Sewell, 1772.
British Patents 542, 1752, and 1971. British Library. London, UK.
Carnot, Sadi. *Reflections on the Motive Power of Fire and on Machines Fitted to Develop That Power*. Translated by R. H. Thurston. Paris: Chez Bachelier Libraire, 1824.
Defoe, Daniel. *A Tour Thro' the Whole Island of Great Britain, Divided into Circuits or Journeys*. 4 vols. 4th ed. London: S. Birt etc., 1748.
Hutton, William. The History of Derby. In D. B. Horn and Mary Ransome, eds. *English Historical Documents, Vol. X, 1714–1783*. New York: Oxford University Press, 1969.
Lemire, Beverly, ed. *The British Cotton Trade, 1660–1815*. London: Pickering & Chatto, 2010.

Marx, Karl, and Friedrich Engels. "Manifesto of the Communist Party." Pamphlet, 1848.

Radcliffe, William. *Origin of the New System of Manufacture, Commonly Called "Power-Loom Weaving."* Stockport: James Lomax, Advertiser-Office, 1828.

Smith, Adam. *An Inquiry into the Nature and Causes of the Wealth of Nations.* London: W. Strahan and T. Cadell, 1776.

Reference Works

Black, John C. *Dictionary of Economics.* 2nd ed. Oxford and New York: Oxford University Press, 2002.

Eltis, David, et al. *The Trans-Atlantic Slave Trade Database.* Atlanta: Emory University. Accessed February 17, 2017. www.slavevoyages.org/.

Encyclopedia Britannica

Heilbrunn Timeline of Art History. New York: Metropolitan Museum of Art, 2000–. https://www.metmuseum.org/toah/. Accessed Oct. 6, 2019.

Historic England Listings

Letters, Samantha. "Gazetteer of Markets and Fairs in England and Wales to 1516." Accessed October 8, 2018. www.history.ac.uk/cmh/gaz/gazweb1 .html.

"Manchester Maps," n.d. Accessed May 15, 2019. http://manchester.publicprofiler .org/beta/.

"Mineral Fact Sheet: Fuller's Earth," British Geological Survey. Accessed December 14, 2015.

Ordnance Survey Maps

Oxford Dictionary of National Biography

Oxford English Dictionary

"Printed Textiles from the 18th Century," Accessed October 5, 2016. www.musee-impression.com/gb/collection/xviii.html, xviii–xix.

Secondary Sources

Adamson, Glenn, Giorgio Riello, and Sarah Teasley, eds. *Global Design History.* London: Routledge, 2011.

Ajmar-Wollheim, Marta, and Luca Molà, "The Global Renaissance: Cross-Cultural Objects in the Early Modern Period." In Glenn Adamson, Giorgio Riello, and Sarah Teasley (eds.), *Global Design History*, 11–20. London: Routledge, 2011.

Alexander, Jennifer Karns. *The Mantra of Efficiency: From Waterwheel to Social Control.* Baltimore: Johns Hopkins University Press, 2008.

Alison, Archibald. *History of Europe, from the fall of Napoleon to the Accession of Louis Napoleon.* Edinburgh: William Blackwood and Sons, 1855.

Allen, Robert C. *The British Industrial Revolution in Global Perspective.* Cambridge and New York: Cambridge University Press, 2009.

"The Industrial Revolution in Miniature: The Spinning Jenny in Britain, France, and India." *Journal of Economic History* 69, no. 4 (December 2009): 418–35.

"The Spinning Jenny: A Fresh Look." *Journal of Economic History* 71, no. 2 (June 2011): 461–64.

"Tracking the Agricultural Revolution in England," *Economic History Review* 52, no. 2 (May 1999): 209–35.

Appleby, Joyce. *The Relentless Revolution: A History of Capitalism*. New York: W. W. Norton & Company, 2010.

Ashton, T. S. *The Industrial Revolution, 1760–1830*. Oxford and New York: Oxford University Press, 1968.

Ashworth, W. J. *The Industrial Revolution: The State, Knowledge and Global Trade*. London: Bloomsbury, 2017.

Aspin, Christopher. "New Evidence on James Hargreaves and the Spinning Jenny." *Textile History* 1, no. 1 (1968): 119–21.

The Water-Spinners. Helmshore: Helmshore Local History Society, 2003.

The Woollen Industry. Buckinghamshire: Shire Publications, 1994.

Aspin, Christopher, and Stanley D. Chapman. *James Hargreaves and the Spinning Jenny*. Preston: Helmshore Local Historical Society, 1964.

Baines, Jr., Edward. *History of the Cotton Manufacture in Great Britain*. London: H. Fisher, R. Fisher, and P. Jackson, 1835.

Baker, Bruce E., and Barbara Hahn. *The Cotton Kings: Capitalism and Corruption in Turn-of-the-Century New York and New Orleans*. New York: Oxford University Press, 2016.

"Cotton." *Essential Civil War Curriculum* (2015). Accessed June 18, 2017. www.essentialcivilwarcurriculum.com/cotton.html.

Barnwell, P. S., Marilyn Palmer, and Malcolm Airs, eds. *The Vernacular Workshop: From Craft to Industry, 1400–1700*. York: Council for British Archaeology, 2004.

Barrie, D. S. "The Liverpool & Manchester Railway: A Centenary of World-Wide Interest." *Railway and Locomotive Historical Society Bulletin* 22 (May 1930): 46–49.

Bartlett, Robert. *England under the Norman and Angevin Kings, 1075–1225*. New York and Oxford: Clarendon Press, 2000.

Beckert, Sven. *Empire of Cotton: A Global History*. New York: Alfred A. Knopf, 2014.

Belfanti, Carlo. "Guilds, Patents, and the Circulation of Technical Knowledge: Northern Italy during the Early Modern Age." *Technology and Culture* 45, no. 3 (July 2004): 569–89.

Bell, Adrian R., Chris Brooks, and Paul R. Dryburgh. *The English Wool Market, c.1230–1327*. Cambridge and New York: Cambridge University Press, 2007.

Bennett, Robert J. *Local Business Voice: The History of Chambers of Commerce in Britain, Ireland and Revolutionary America 1760–2011*. London and New York: Oxford University Press, 2011.

Benson, Anna P. *Textile Machines*. Princes Risborough: Shire Publications, 1983.

Bentley, Jerry H., and Sanjay Subrahmanyam, eds. *The Cambridge World History: Volume 6, The Construction of a Global World, 1400–1800 CE, Part 1, Foundations*. Cambridge: Cambridge University Press, 2015.

Berg, Maxine. *The Age of Manufactures: Industry, Innovation, and Work in Britain, 1700–1820*. Oxford: Basil Blackwell, in association with Fontana, 1985.

The Machinery Question and the Making of Political Economy 1815–1848. Rev. ed. Cambridge and New York: Cambridge University Press, 1980.

Berg, Maxine, and Pat Hudson. "Rehabilitating the Industrial Revolution." *Economic History Review*, New Series, 45, no. 1 (February 1992): 24–50.

Bianchi, Carlotta, Fabio Cani, et al. *Guide to the Educational Silk Museum of Como*. Como: Museo didattico della Seta, 2004.

Biggs, Lindy. *The Rational Factory: Architecture, Technology and Work in America's Age of Mass Production*. Baltimore: Johns Hopkins University Press, 2003.

Bijker, Wiebe E., and Annapurna Mamidipudi. "Mobilising Discourses: Handloom as Sustainable Socio-Technology." *Economic and Political Weekly* XLVII, no. 25 (June 23, 2012): 41–51.

Bijker, Wiebe E., Thomas Parke Hughes, and Trevor Pinch, eds. *The Social Construction of Technological Systems: New Directions in the Sociology and History of Technology*. Anniversary ed. Cambridge, MA: MIT Press, 2012.

Blaszczyk, Regina Lee. *Fashionability: Abraham Moon and the Creation of British Cloth for the Global Market*. Manchester: Manchester University Press, 2017.

Imagining Consumers: Design and Innovation from Wedgwood to Corning. Baltimore and London: Johns Hopkins University Press, 2000.

Bogost, Ian. "What Is Object-Oriented Ontology? A Definition for Ordinary Folk." Blogpost, December 8, 2009. Accessed July 7, 2016. http://bogost .com/writing/blog/what_is_objectoriented_ontolog/.

Bottomley, Sean. "The British Patent System during the Industrial Revolution, 1700–1852." Ph.D. Diss., University of Cambridge, 2013.

Bowden, Witt. "The Influence of the Manufacturers on Some of the Early Policies of William Pitt." *American Historical Review* 29, no. 4 (July 1924): 655–74.

Bowen, H. V. *The Business of Empire: The East India Company and Imperial Britain, 1756–1833*. Cambridge: Cambridge University Press, 2006.

Bowen, H. V., Margarette Lincoln, and Nigel Rigby. *The Worlds of the East India Company*. Suffolk: Boydell & Brewer, 2002.

Boyle, James E. *Cotton and the New Orleans Cotton Exchange: A Century of Commercial Evolution*. Garden City, NY: Doubleday, Doran & Co., Inc., 1934.

Bridbury, A. R. *Medieval English Clothmaking: An Economic Survey*. London: Heinemann Educational Books and the Pasold Research Fund, 1982.

Britnell, R. H. "The Towns of England and Northern Italy in the Early Fourteenth Century." *Economic History Review*, New Series, 44, no. 1 (February 1991): 21–35.

Broadberry, Stephen, and Bishnupriya Gupta. "Cotton Textiles and the Great Divergence: Lancashire, India, and Shifting Competitive Advantage, 1600–1850." Centre for Economic Policy Research Discussion Paper No. 5183, 25 Aug. 2005.

Brown, David. "'Persons of Infamous Character' or 'an Honest, Industrious and Useful Description of People'? The Textile Pedlars of Alstonfield and the Role of Peddling in Industrialization." *Textile History* 31, no. 1 (2000): 1–26.

Brown, John K. *The Baldwin Locomotive Works, 1831–1915: A Study in American Industrial Practice.* Baltimore: Johns Hopkins University Press, 2001.

Bruchey, Stuart. *Cotton and the Growth of the American Economy, 1790–1860: Sources and Readings.* New York and Chicago: Harcourt, Brace & World, Inc., 1967.

Bryden, D. J. "James Watt, Merchant: The Glasgow Years, 1754–1774." In Denis Smith (ed.), *Perceptions of Great Engineers: Fact and Fantasy,* 9–22. London and Liverpool: Science Museum for the Newcomen Society, National Museums and Galleries on Merseyside and the University of Liverpool, 1994.

Bryson, Bill. *Notes from a Small Island.* London: Doubleday, 1995.

Burnley, James. *Sir Titus Salt, and George Moore.* London: Cassell & Co., 1885.

Bythell, Duncan. *The Handloom Weavers: A Study in the English Cotton Industry During the Industrial Revolution.* Cambridge: Cambridge University Press, 1969.

Cain, P. J., and A. G. Hopkins. *British Imperialism: 1688–2015.* 3rd ed. Abingdon and New York: Routledge, 2016.

"Gentlemanly Capitalism and British Overseas Expansion I: The Old Colonial System, 1688–1850." *Economic History Review,* New Series, 39, no. 4 (November 1986): 501–25.

"Gentlemanly Capitalism and British Expansion Overseas II: New Imperialism, 1850–1945." *Economic History Review,* New Series, 40, no. 1 (February 1987): 1–26.

Cannadine, David. "The Present and the Past in the English Industrial Revolution, 1880–1980." *Past and Present* 103, no. 1 (May 1984): 131–72.

Cardwell, Donald S. *From Watts to Clausius: The Rise of Thermodynamics in the Early Industrial Age.* Ithaca, NY: Cornell University Press, 1971.

Carnegie, Andrew. *Autobiography of Andrew Carnegie, with Illustrations.* London: Constable & Co., 1920.

Carson, Cary, and Ronald Hoffman, eds. *Of Consuming Interests: The Style of Life in the Eighteenth Century.* Charlottesville and London: University Press of Virginia, 1994.

Carus-Wilson, E. M. "An Industrial Revolution of the Thirteenth Century." *Economic History Review* 11, no. 1 (October 1941): 39–60.

Catling, Harold. "The Development of the Spinning Mule." *Textile History* 9, no. 1 (1978): 35–57.

The Spinning Mule. Newton Abbot: David & Charles, 1970.

Chapman, Stanley. *The Early Factory Masters: The Transition to the Factory System in the Midlands Textile Industry.* Newton Abbot: David & Charles, 1967.

Merchant Enterprise in Britain from the Industrial Revolution to World War I. Cambridge and New York: Cambridge University Press, 1992.

The Rise of Merchant Banking, 10–11, cited in Boyd Hilton, *A Mad, Bad, and Dangerous People?* Oxford: Clarendon Press, 2006, p. 23.

Chartres, John, and Katrina Honeyman, eds. *Leeds City Business, 1893–1993: Essays Marking the Centenary of the Incorporation.* Leeds: Leeds University Press, 1993.

Chase, Malcolm. *Chartism: A New History.* Manchester and New York: Manchester University Press, 2007.

   1820: Disorder and Stability in the United Kingdom. Manchester and New York: Manchester University Press, 2013.

Chaudhuri, K. N. *The English East India Company: The Study of an Early Joint-Stock Company 1600–1640.* London: Frank Cass & Co., 1965.

Checkland, S. G. "American versus West Indian Traders in Liverpool, 1793–1815." *Journal of Economic History* 18, no. 2 (June 1958): 141–60.

Clapham, J. H. *An Economic History of Modern Britain.* 3 vols. Cambridge: Cambridge University Press, 1926.

Clark, Gregory. *A Farewell to Alms: A Brief Economic History of the World.* Princeton and Oxford: Princeton University Press, 2007.

   "The Long March of History: Farm Wages, Population, and Economic Growth, England 1209–1869." *Economic History Review,* New Series, 60, no. 1 (February 2007): 97–135.

Clark, Gregory, Kevin Hjortshøj O'Rourke, and Alan M. Taylor. "The Growing Dependence of Britain on Trade during the Industrial Revolution." University of Oxford, Discussion Papers in Economic and Social History, No. 126 (March 2014). www.economics.ox.ac.uk/materials/papers/13258/clark-et-al-new-world.pdf.

Clark, Peter, and Paul Slack. *English Towns in Transition 1500–1700.* London: Oxford University Press, 1976.

Cooke, Anthony. *The Rise and Fall of the Scottish Cotton Industry, 1778–1914: "The Secret Spring."* Manchester and New York: Manchester University Press, 2010.

   "The Scottish Cotton Masters, 1780–1914." *Textile History* 40, no. 1 (2009): 29–50.

Cookson, Gillian. "A City in Search of Yarn: The Journal of Edward Taylor of Norwich, 1817." *Textile History* 37, no. 1 (May 2006): 38–51.

   The Age of Machinery: Engineering the Industrial Revolution, 1770–1850. Woodbridge: Boydell Press, 2018.

   Victoria County History: A History of the County of Durham. Vol. 4. Woodbridge: Boydell and Brewer for the Institute of Historical Research, 2005.

   "The West Yorkshire Textile Engineering Industry, 1780–1850." Ph.D. Diss., University of York, 1994.

Crafts, N. F. R. *British Economic Growth during the Industrial Revolution.* New York: The Clarendon Press, Oxford University Press, 1985.

   "British Economic Growth, 1700–1850: Some Difficulties of Interpretation." *Explorations in Economic History* 24 (1987): 245–68.

Crump, William B., ed. *The Leeds Woollen Industry 1780–1820.* Leeds: Thoresby Society, 1931.

Curtin, Philip D. *The Atlantic Slave Trade: A Census.* Madison: University of Wisconsin Press, 1972.

ed. *The Rise and Fall of the Plantation Complex: Essays in Atlantic History*. 2nd. ed. Cambridge and New York: Cambridge University Press, 1998.

Dalzell, Robert F. *Enterprising Elite: The Boston Associates and the World They Made*. Cambridge, MA and London: Harvard University Press, 1987.

Daunton, Martin. *Progress and Poverty: An Economic and Social History of Britain.* Oxford: Oxford University Press, 1995.

Davenport, Elsie G. *Your Handspinning*. Pacific Grove, CA: Select Books, 1964.

Davies, Jeremy. *The Birth of the Anthropocene*. Berkeley: University of California Press, 2016.

Delve, Janet. "Jacques Vaucanson: 'Mechanic of Genius'." *IEEE Annals of the History of Computing* 29, no. 4 (Oct.-Dec. 2007): 94–97.

"Joseph Marie Jacquard: Inventor of the Jacquard Loom." *IEEE Annals of the History of Computing* 29, no. 4 (Oct.-Dec. 2007): 98–102.

Dickenson, Michael. "The West Riding Woollen and Worsted Industries, 1689–1770: An Analysis of Probate Inventories and Insurance Policies." Ph.D. Diss., University of Nottingham, 1974.

Dickson, A., and W. Speirs. "Changes in Class Structure in Paisley, 1750–1845." *Scottish Historical Review* 59, no. 167, Part 1 (April 1980): 54–72.

Dickson, Tony, and Tony Clarke. "Social Concern and Social Control in Nineteenth Century Scotland: Paisley 1841–1843." *Scottish Historical Review* 65, no. 179, Part 1 (April 1986): 48–60.

Doda, Hilary. "'Saide Monstrous Hose': Compliance, Transgression and English Sumptuary Law to 1533." *Textile History* 45, no. 2 (2014): 171–91.

Eacott, Jonathan. *Selling Empire: India in the Making of Britain and America, 1600–1830*. Williamsburg and Chapel Hill: Omohundro Institute of Early American History and Culture and University of North Carolina Press, 2016.

Eastwood, David. *Government and Community in the English Provinces, 1700–1870*. New York: St. Martin's Press, 1997.

Edgerton, David. "Innovation, Technology, or History: What Is the Historiography of Technology About." *Technology and Culture* 51, no. 3 (July 2010): 680–97.

The Shock of the Old: Technology and Global History since 1900. Oxford and New York: Oxford University Press, 2007.

Edmundson, William. *The Nitrate King: A Biography of "Colonel" John Thomas North*. New York: Palgrave Macmillan, 2011.

Ellison, Thomas. *The Cotton Trade of Great Britain. Including a History of the Liverpool Cotton Market and of the Liverpool Cotton Brokers' Association*. London: Effingham Wilson, Royal Exchange, 1886.

Ellul, Jacques. *The Technological Society*. New York: Alfred A. Knopf, 1964.

Engels, Frederick. *The Condition of the Working Class in England in 1844, Complete and Uncensored*. Translated by Florence Kelley Wischnewetzky. Springfield, MA: Seven Treasures Publications, 2009. Orig. pub. 1844.

English, Walter. "A Study of the Driving Mechanisms in the Early Circular Throwing Machines." *Textile History* 2, no. 1 (1971): 65–75.

"A Technical Assessment of Lewis Paul's Spinning Machine." *Textile History* 4, no. 1 (1973): 68–83.

Epstein, S. R. "Craft Guilds, Apprenticeship, and Technological Change in Preindustrial Europe." *Journal of Economic History* 58, no. 3 (September 1998): 684–713.

"Property Rights to Technical Knowledge in Premodern Europe, 1300–1800." *American Economic Review* 94, no. 2 (May 2004): 382–87.

Epstein, S. R., and Maarten Prak, eds. *Guilds, Innovation and the European Economy, 1400–1800.* Cambridge and New York: Cambridge University Press, 2008.

Fairbairn, William, and Thomas Baines. *The Rise and Progress of Manufactures and Commerce and of Civil and Mechanical Engineering in Lancashire and Cheshire.* Glasgow: W. Mackenzie, 1869.

Farnie, Douglas A. *The English Cotton Industry and the World Market 1815–1896.* Oxford: Oxford University Press, 1979.

Farnie, Douglas A., and David Jeremy, eds. *The Fibre That Changed the World: The Cotton Industry in International Perspective, 1600–1990s.* Oxford: Oxford University Press, 2004.

Ferreira, Maria João. "Asian Textiles in the Carreira Da Índia: Portuguese Trade, Consumption and Taste, 1500–1700." *Textile History* 46, no. 2 (2015): 147–68.

Firth, Gary. *Bradford and the Industrial Revolution: An Economic History, 1760–1840.* Halifax, UK: Ryburn Publishing, 1990.

Fitton, R. S., and A. P. Wadsworth. *The Strutts and the Arkwrights, 1758–1830: A Study of the Early Factory System.* Manchester: Manchester University Press, 1958.

Foner, Laura. *Slavery in the New World: A Reader in Comparative History.* New York: Prentice-Hall, 1969.

Fox, Celina. *The Arts of Industry in the Age of Enlightenment.* New Haven: Yale University Press, 2010.

Fox, Robert, ed. *Technological Change: Methods and Themes in the History of Technology.* Amsterdam: Harwood Academic Publishers, 1996.

"Watt's Expansive Principle in the Work of Sadi Carnot and Nicolas Clément." *Notes and Records of the Royal Society of London* 24, no. 2 (April 1970): 233–53.

Fox-Genovese, Elizabeth. *Within the Plantation Household: Black and White Women of the Old South.* New ed. Chapel Hill: University of North Carolina Press, 1988.

Frank, Andre Gunder. "A Plea for World Systems History." *Journal of World History* 2, no. 1 (Spring 1991): 1–28.

ReOrient: Global Economy in the Asian Age. Berkeley and Los Angeles: University of California Press, 1998.

Freifeld, Mary. "Technological Change and the 'Self-Acting' Mule: A Study of Skill and the Sexual Division of Labor." *Social History* 11, no. 3 (October 1986): 319–43.

French, Gilbert J. *The Life and Times of Samuel Crompton, Inventor of the Spinning Machine Called the Mule, with an Appendix of Original Documents.* 2nd ed. Manchester and London: Thomas Dinham & Co.; Simpkin, Marshall & Co., 1860.

Gauci, Perry. *The Politics of Trade: The Overseas Merchants in State and Society, 1660–1720.* Oxford: Oxford University Press, 2001.

Gerritsen, Anne. "Global Design in Jingdezhen: Local Production and Global Connections." In Glenn Adamson, Giorgio Riello, and Sarah Teasley (eds.), *Global Design History,* 25–33. London: Routledge, 2011.

Giles, Colum, and Ian H. Goodall. *Yorkshire Textile Mills 1770–1930: The Buildings of the Yorkshire Textile Industry, 1770–1930.* London: H.M. Stationery Office, 1992.

Gillow, John, and Nicholas Barnard. *Indian Textiles.* London: Thames and Hudson, 2008.

Goldstone, Jack A. "Efflorescences and Economic Growth in World History: Rethinking the 'Rise of the West' and the Industrial Revolution." *Journal of World History* 13, no. 2 (Fall 2002): 323–89.

Goodwin, Albert. *The Friends of Liberty: The English Democratic Movement in the Age of the French Revolution.* London and New York: Routledge, 2016. Orig. pub. 1979.

Grafe, Regina. "Review of Epstein and Prak, *Guilds, Innovation, and the European Economy.*" *Journal of Interdisciplinary History* 40, no. 1 (Summer 2009): 78–82.

Gragnolati, Ugo, Daniele Moschella, and Emanuele Pugliese. "The Spinning Jenny and the Industrial Revolution: A Reappraisal." *Journal of Economic History* 71, no. 2 (June 2011): 455–60.

Griffin, Emma. "A Conundrum Resolved? Rethinking Courtship, Marriage, and Population Growth in Eighteenth-Century England." *Past and Present,* no. 215 (May 2012): 125–64.

"The 'Industrial Revolution': Interpretations from 1830 to the Present." School of History, University of East Anglia, 2013. Accessed May 2, 2017. www .uea.ac.uk/documents/1006128/1446434/Emma+Griffin+industrialrevolu tion.pdf/816bcd4c-ac9b-4700-aae4-ee6767d4f04a.

Liberty's Dawn: A People's History of the Industrial Revolution. New Haven: Yale University Press, 2013.

A Short History of the British Industrial Revolution. New York: Palgrave Macmillan, 2010.

Griffiths, Trevor, Philip A. Hunt, and Patrick K. O'Brien. "Inventive Activity in the British Textile Industry, 1700–1800." *Journal of Economic History* 52, no. 4 (December 1992): 881–906.

"Scottish, Irish, and Imperial Connections: Parliament, the Three Kingdoms, and the Mechanization of Cotton Spinning in Eighteenth-Century Britain." *Economic History Review,* New Series, 61, no. 3 (August 2008): 625–50.

Guy, John. "'One Thing Leads to Another': Indian Textiles and the Early Globalization of Style." In Amelia Peck (ed.), *Interwoven Globe: The Worldwide Textile Trade, 1500–1800,* 12–27. London: Thames and Hudson, 2013.

Haase, Donald. "The Sleeping Script: Memory and Forgetting in Grimms' Romantic Fairy Tale (KHM 50)." *Merveilles & Contes* 4, no. 2 (December 1990): 166–76.

Habakkuk, H. J. *American and British Technology in the Nineteenth Century: The Search for Labour Saving Inventions.* Cambridge: Cambridge University Press, 1967.

Hafter, Daryl M. *Women at Work in Preindustrial France.* University Park: Pennsylvania State University Press, 2007.

Hahn, Barbara. *Making Tobacco Bright: Creating an American Commodity, 1617–1937.* Baltimore and New York: Johns Hopkins University Press, 2011.

"The Social in the Machine: How and Why the History of Technology Looks Beyond the Object." *Perspectives on History: The Newsmagazine of the American Historical Association* (March 2014): 30–31.

Hall, Robert G. "Tyranny, Work and Politics: The 1818 Strike Wave in the English Cotton District." *International Review of Social History* 34, no. 3 (1989): 433–70.

Hallas, Christine. "Cottage and Mill: The Textile Industry in Wensleydale and Swaledale." *Textile History* 21, no. 2 (1990): 203–21.

Hammond, J. L., and Barbara Hammond. *The Town Labourer, 1760–1832: The New Civilisation.* 4th impression. London: Longmans, Green, and Co., 1919.

Hardesty, Jared Ross. *Unfreedom: Slavery and Dependence in Eighteenth-Century Boston.* New York: New York University Press, 2016.

Harley, C. K. "British Industrialisation before 1841: Evidence of Slower Growth during the Industrial Revolution." *Journal of Economic History* 42, no. 2 (June 1982): 267–89.

Harman, Graham. *Tool-Being: Heidegger and the Metaphysics of Objects.* Peru, IL: Open Court, 2002.

Harris, Ron. *Industrializing English Law: Entrepreneurship and Business Organization, 1720–1844.* Cambridge and New York: Cambridge University Press, 2010.

Hathaway, Ian. "How Big Is the Tech Sector?" Blogpost. Accessed July 25, 2017. www.ianhathaway.org/blog/2017/5/31/how-big-is-the-tech-sector.

Hayman, Richard, and Wendy Horton. *Ironbridge: History and Guide.* Stroud, Gloucestershire: History Press, 2009.

Heaton, Herbert. "Benjamin Gott and the Industrial Revolution in Yorkshire." *Economic History Review* 3, no. 1 (January 1931): 45–66.

"The Leeds White Cloth Hall." *The Publications of the Thoresby Society* 22 (1915): 131–71.

The Yorkshire Woollen and Worsted Industries, from the Earliest Times up to the Industrial Revolution. 2nd ed. Oxford: Clarendon Press, 1965.

Hemingway, Ernest. *The Sun Also Rises.* New York: Scribner, 1954.

Hilaire-Pérez, Liliane, and Catherine Verna. "Dissemination of Technical Knowledge in the Middle Ages and the Early Modern Era: New Approaches and Methodological Issues." *Technology and Culture* 47, no. 3 (July 2006): 536–65.

Hills, Richard L. *Power from Steam: A History of the Stationary Steam Engine.* Cambridge: Cambridge University Press, 1989.

Power in the Industrial Revolution. Manchester: Manchester University Press, 1970.

Hilton, Boyd. *A Mad, Bad, and Dangerous People?: England 1783–1846*. The New Oxford History of England. Oxford: Clarendon Press, 2006.

Hirshler, Eric E. "Medieval Economic Competition." *Journal of Economic History* 14, no. 1 (Winter 1954): 52–58.

Hobsbawm, Eric J. *The Age of Revolution: 1789–1848*. London: Weidenfeld and Nicholson, 1962.

"The Machine Breakers." *Past and Present* 1 (February 1952): 57–70.

Hodge, Eric C. "'Their Palms Were Crossed with Silver': The Payment of Workers in Early Textile Factories 1780–1830." *Textile History* 40, no. 2 (2009): 229–37.

Hoffman, Philip T. *Why Did Europe Conquer the World?* Princeton: Princeton University Press, 2015.

Holt, Robin, and Andrew Popp. "Josiah Wedgwood, Manufacturing and Craft." *Journal of Design History* 29, no. 2 (January 2016): 99–119.

Honeyman, Katrina. *Child Workers in England, 1780–1820: Parish Apprentices and the Making of the Early Industrial Labour Force*. Aldershot, England, and Burlington, VT: Ashgate Publishing, 2007.

Well Suited: A History of the Leeds Clothing Industry, 1850–1990. Oxford: Oxford University Press, 2000.

Horn, Jeff. "Machine-Breaking in England and France during the Age of Revolution." *Labour* 55 (Spring 2005): 143–67.

Horn, Jeffrey, Leonard N. Rosenband, and Merritt Roe Smith, eds. *Reconceptualizing the Industrial Revolution*. Cambridge, MA: MIT Press, 2010.

Horrell, Sara, and Jane Humphries. "Women's Labour Force Participation and the Transition to the Male-Breadwinner Family, 1790–1865." *Economic History Review*, New Series, 48, no. 1 (February 1995): 89–117.

Huang, Angela Ling, and Carsten Jahnke, eds. *Textiles and the Medieval Economy: Production, Trade, and Consumption of Textiles, 8th–16th Centuries*. Oxford and Philadelphia: Oxbow Books, 2015.

Hudson, Pat. *The Genesis of Industrial Capital: A Study of the West Riding Wool Textile Industry, c. 1750–1850*. Cambridge: Cambridge University Press, 1986.

Hughes, Thomas P. *Networks of Power: Electrification in Western Society, 1880–1930*. Baltimore and London: Johns Hopkins University Press, 1983.

Rescuing Prometheus: Four Monumental Projects That Changed Our World. New York: Vintage Books, 2000.

Humphries, Jane. *Childhood and Child Labour in the British Industrial Revolution*. Cambridge and New York: Cambridge University Press, 2010.

Humphries, Jane, and Benjamin Schneider. "Spinning the Industrial Revolution." *Economic History Review* 72, no. 1 (2019): 126–55.

Humphries, Jane, and Jacob Weisdorf. "The Wages of Women in England, 1260–1850." *Journal of Economic History* 75, no. 2 (June 2015): 405–47.

Hunt, Edwin S., and James Murray. *A History of Business in Medieval Europe, 1200–1550*. Cambridge: Cambridge University Press, 1999.

Hurst, Derek. *Sheep in the Cotswolds: The Medieval Wool Trade.* Stroud, Gloucestershire: History Press, 2005.

Hyde, Francis E. *Liverpool and the Mersey: An Economic History of a Port 1700–1970.* Newton Abbot: David & Charles, 1971.

Hyde, Francis E., Bradbury B. Parkinson, and Sheila Marriner. "The Cotton Broker and the Rise of the Liverpool Cotton Market." *Economic History Review*, New Series, 8, no. 1 (1955): 75–83.

Inikori, Joseph E. *Africans and the Industrial Revolution in England: A Study in International Trade and Economic Development.* Cambridge and New York: Cambridge University Press, 2002.

Irwin, Douglas A. "Political Economy and Peel's Repeal of the Corn Laws." *Economics and Politics* 1, no. 1 (Spring 1989): 41–59.

Isaacson, Walter. *The Innovators: How a Group of Hackers, Geniuses, and Geeks Created the Digital Revolution.* New York: Simon & Schuster, 2014.

Steve Jobs. New York: Simon & Schuster, 2011.

Jackson, Kenneth C. "The Room and Power System in the Cotton Weaving Industry of North-East Lancashire and West Craven." *Textile History* 35, no. 1 (2004): 58–89.

Jacob, Margaret C. *The First Knowledge Economy: Human Capital and the European Economy, 1750–1850.* Cambridge and New York: Cambridge University Press, 2014.

Scientific Culture and the Making of the Industrial West. New York and Oxford: Oxford University Press, 1997.

James, Kevin J. *Handloom Weavers in Ulster's Linen Industry, 1815–1914.* Dublin: Four Courts Press, 2007.

Jenkins, D. T. *The Cambridge History of Western Textiles.* Vol. 1. Cambridge: Cambridge University Press, 2003.

Jenkins, J. Geraint, ed. *The Wool Textile Industry in Great Britain.* London and Boston: Routledge & Kegan Paul, 1972.

Jeremy, David J. "British Textile Technology Transmission to the United States: The Philadelphia Region Experience, 1770–1820." *Business History Review* 47, no. 1 (Spring 1973): 24–52.

"Innovation in American Textile Technology during the Early Nineteenth Century." *Technology and Culture* 14, no. 1 (January 1973): 40–76.

Transatlantic Industrial Revolution: The Diffusion of Textile Technologies between Britain and America, 1790–1830. Cambridge, MA: MIT Press, 1981.

Jones, Eric. *The European Miracle: Environments, Economies and Geopolitics in the History of Europe and Asia.* Cambridge and New York: Cambridge University Press, 2003.

Jones, Peter M. *Agricultural Enlightenment: Knowledge, Technology, and Nature, 1750–1840.* New York: Oxford University Press, 2016.

Industrial Enlightenment: Science, Technology, and Culture in Birmingham and the West Midlands, 1760–1820. Manchester and New York: Manchester University Press, 2008.

Kander, Astrid, Paolo Malanima, and Paul Warde. *Power to the People: Energy in Europe over the Last Five Centuries.* Princeton and Oxford: Princeton University Press, 2013.

Karpinski, Caroline. "Kashmir to Paisley." *Metropolitan Museum of Art Bulletin,* New Series, 22, no. 3 (November 1963): 116–23.

Kelly, Morgan, Joel Mokyr, and Cormac Ó Gráda. "Precocious Albion: A New Interpretation of the British Industrial Revolution." *Annual Review of Economics* 6 (2014): 363–89.

"Roots of the Industrial Revolution." UCD Centre for Economic Research Working Paper Series, WP2015/24, October 2015. Accessed May 15, 2019. http://researchrepository.ucd.ie/handle/10197/7183.

Kerker, Milton. "Science and the Steam Engine." *Technology and Culture* 2, no. 4 (Autumn 1961): 381–90.

Kidd, Alan, and Terry Wyke, eds. *Manchester: Making the Modern City.* Liverpool: Liverpool University Press, 2016.

Kindleberger, Charles P. *World Economic Primacy: 1500–1990.* Oxford and New York: Oxford University Press, 1996.

King, Steven, and Geoffrey Timmins. *Making Sense of the Industrial Revolution: English Economy and Society 1700–1850.* Manchester and New York: Manchester University Press, 2001.

Knight, Frida. *The Strange Case of Thomas Walker: Ten Years in the Life of a Manchester Radical.* London: Lawrence & Wishart, 1957.

Komlos, John. "Nutrition, Population Growth, and the Industrial Revolution in England." *Social Science History* 14, no. 1 (Spring 1990): 69–91.

Kriedte, Peter, Hans Medick, and Jürgen Schlumbohm. *Industrialization before Industrialization: Rural Industry in the Genesis of Capitalism.* Translated by Beatrice Schempp. Cambridge and London: Cambridge University Press and Paris, Editions de la Maison des Sciences de l'Homme, 1981.

Kriedte, Peter, "Decline of Proto-Industrialization, Pauperism, and the Sharpening of the Contrast between City and Countryside." In Peter Kriedte, Hans Medick, and Jürgen Schlumbohm (eds.), *Industrialization before Industrialization: Rural Industry in the Genesis of Capitalism,* 154–60. Cambridge and London: Cambridge University Press and Paris, Editions de la Maison des Sciences de l'Homme, 1981.

Kulke, Hermann, and Dietmar Rothermund. *A History of India.* 6th ed. New York: Routledge, 2016.

Landes, David S. *Unbound Prometheus: Technological Change and Industrial Development in Western Europe from 1750 to the Present.* Cambridge: Cambridge University Press, 1969.

The Wealth and Poverty of Nations: Why Some Are So Rich and Some So Poor. New York: W. W. Norton & Company, 1998.

Langdon, John, and James Masschaele. "Commercial Activity and Population Growth in Medieval England." *Past and Present,* no. 190 (February 2006): 35–81.

Latour, Bruno. *Reassembling the Social: An Introduction to Actor-Network-Theory.* Oxford and New York: Oxford University Press, 2005.

Law, John, and John Hassard, eds. *Actor Network Theory and After.* Oxford: Wiley-Blackwell, 1999.

Lawson, Philip. *The East India Company: A History.* London and New York: Longman, 1987.

Lazonick, William. *Competitive Advantage on the Shop Floor*. Cambridge, MA: Harvard University Press, 1990.

"Industrial Relations and Technical Change: The Case of the Self-Acting Mule." *Cambridge Journal of Economics* 3, no. 3 (September 1979): 231–62.

Leadbeater, Eliza. *Spinning and Spinning Wheels*. Buckinghamshire: Shire Publications, 1995.

Leavitt, Thomas W. "Fashion, Commerce and Technology in the Nineteenth Century: The Shawl Trade." *Textile History* 3, no. 1 (1972): 51–63.

Lee, John S. *The Medieval Clothier*. Woodbridge: Boydell Press, 2018.

Lemire, Beverly, ed. *The British Cotton Trade, 1660–1815*. London: Pickering & Chatto, 2010.

"Consumerism in Preindustrial and Early Industrial England: The Trade in Secondhand Clothes." *Journal of British Studies* 27, no. 1 (January 1988): 1–24.

Cotton. Oxford: Berg, 2011.

Fashion's Favorite: The Cotton Trade and the Consumer in Britain, 1660–1800. Oxford and New York: Oxford University Press, 1991.

"'A Good Stock of Cloaths': The Changing Market for Cotton Clothing in Britain, 1750–1800." *Textile History* 22, no. 2 (1991): 311–28.

Lemire, Beverly, and Giorgio Riello. "East & West: Textiles and Fashion in Early Modern Europe." *Journal of Social History* 41, no. 4 (Summer 2008): 887–916.

Long, Pamela O. *Openness, Secrecy, and Authorship: Technical Arts and the Culture of Knowledge from Antiquity to the Renaissance*. Baltimore: Johns Hopkins University Press, 2001.

"Trading Zones in Early Modern Europe." *ISIS* 106, no. 4 (2015): 840–47.

Lovejoy, Paul E. *Transformations in Slavery: A History of Slavery in Africa*. 3rd ed. New York: Cambridge University Press, 2012.

Lucas, Adam Robert. "Industrial Milling in the Ancient and Medieval Worlds: A Survey of the Evidence for an Industrial Revolution in Medieval Europe." *Technology and Culture* 46, no. 1 (January 2005): 1–30.

Wind, Water, Work: Ancient and Medieval Milling Technology. Leiden and Boston: Brill, 2006.

Lucassen, Jan, Tine De Moor, and Jan Luiten van Zanden, eds. *The Return of the Guilds*. *International Review of Social History* 53, supplement 16 (2008).

Lyons, Agnes M. M. "The Textile Fabrics of India and Huddersfield Cloth Industry." *Textile History* 27, no. 2 (1996): 172–94.

Ma, Debin. *Textiles in the Pacific, 1500–1900*. Oxon: Ashgate Press, 2005.

MacKeith, Margaret. *The History and Conservation of Shopping Arcades*. London and New York: Mansell Publishing, 1986.

MacKenzie, Donald. "Marx and the Machine." *Technology and Culture* 25, no. 3 (July 1984): 473–502.

MacLeod, Christine. *Heroes of Invention: Technology, Liberalism, and British Identity, 1750–1914*. Cambridge and New York: Cambridge University Press, 2007.

Inventing the Industrial Revolution: The English Patent System, 1660–1800. Cambridge and New York: Cambridge University Press, 1988.

Malm, Andreas. *Fossil Capital: The Rise of Steam Power and the Roots of Global Warming*. London and New York: Verso, 2016.

Mann, Julia De L."The Textile Industry: Machinery for Cotton, Flax, Wool, 1760–1850." In Charles Singer, E. J. Holmyard, A. R. Hall, and Trevor Williams (eds.), *A History of Technology*, 5 vols., 4: 277–307. New York and London: Oxford University Press, 1958.

Masschaele, James. *Peasants, Merchants, and Markets: Inland Trade in Medieval England, 1150–1350*. New York: St. Martin's Press, 1997.

Mather, Ruth. "The Impact of the French Revolution in Britain." British Library, 14 May 2014. Accessed May 15, 2019. www.bl.uk/romantics-and-victorians /articles/the-impact-of-the-french-revolution-in-britain.

Maw, Peter. *Transport and the Industrial City: Manchester and the Canal Age, 1750–1850*. Manchester and New York: Manchester University Press, 2013.

Maw, Peter, Terry Wyke, and Alan Kidd. "Canals, Rivers, and the Industrial City: Manchester's Industrial Waterfront, 1790–1850." *Economic History Review* 65, no. 4 (November 2012): 1495–1523.

Mccants, Anne E. C. "Exotic Goods, Popular Consumption, and the Standard of Living: Thinking about Globalization in the Early Modern World." *Journal of World History* 18, no. 4 (2007): 433–62.

McCusker, John J., and Russell R. Menard. *The Economy of British America, 1607–1789, with Supplementary Bibliography*. Chapel Hill and London: Omohundro Institute of Early American History and Culture and University of North Carolina Press, 1991. Orig. pub. 1985.

Meisenzahl, Ralf, and Joel Mokyr. "The Rate and Direction of Invention in the British Industrial Revolution: Incentives and Institutions." NBER Working Paper No. 16993, April 2011.

Menon, Meena, and Uzramma. *A Frayed History: The Journey of Cotton in India*. New Delhi: Oxford University Press, 2017.

Mingay, G. E., ed. *The Agricultural Revolution: Changes in Agriculture, 1650–1880*. London: Adam & Charles Black, 1977.

Mintz, Sidney W. *Sweetness and Power: The Place of Sugar in Modern History*. New York: Penguin Books, 1986.

Misa, Thomas J. *Leonardo to the Internet: Technology and Culture from the Renaissance to the Present*. Baltimore and New York: Johns Hopkins University Press, 2004.

Mohanty, Gail Fowler. *Labors and Laborers of the Loom: Mechanization and Handloom Weavers, 1780–1840*. New York and London: Routledge, 2006.

Mokyr, Joel, ed. *The British Industrial Revolution: An Economic Perspective*. Boulder: Westview Press, 1993.

A Culture of Growth: The Origins of the Modern Economy. Princeton: Princeton University Press, 2016.

The Enlightened Economy: An Economic History of Britain 1700–1850. New Haven: Yale University Press, 2012.

The Lever of Riches: Technological Creativity and Economic Progress. New York: Oxford University Press, 1992.

Moore, Jason W. "The Capitalocene, Part I: On the Nature and Origins of Our Ecological Crisis." *Journal of Peasant Studies* 44, no. 3 (2017): 594–630.

Morgan, Carol E. "Women, Work and Consciousness in the Mid-Nineteenth-Century English Cotton Industry." *Social History* 17, no. 1 (January 1992): 23–41.

Morrell, J. B. "The Early Yorkshire Geological and Polytechnic Society: A Reconsideration." *Annals of Science* 45 (1988): 153–67.

Mosely, Stephen. *The Chimney of the World: A History of Smoke Pollution in Victorian and Edwardian Manchester*. Abingdon and New York: Routledge, 2008.

Muldrew, Craig. *Food, Energy, and the Creation of Industriousness: Work and Material Culture in Agrarian England, 1550–1780*. Cambridge and New York: Cambridge University Press, 2011.

Mumford, Lewis. *Technics and Civilization*. New York: Harcourt, Brace and Co., 1934.

Munro, John H. "Medieval Woollens: Textiles, Textile Technology and Industrial Organisation, c. 800–1500." In D. T. Jenkins (ed.), *The Cambridge History of Western Textiles*, Vol. 1, 181–227. Cambridge: Cambridge University Press, 2003.

Murray, Norman. *The Scottish Handloom Weavers: A Social History*. Edinburgh: John Donald Publishers, 1978.

Musson, A. E., and E. Robinson. "The Early Growth of Steam Power." *Economic History Review*, New Series, 11, no. 3 (1959): 418–39.

Nardinelli, Clark. "Child Labor and the Factory Acts." *Journal of Economic History* 40, no. 4 (December 1980): 739–55.

  Child Labor and the Industrial Revolution. Bloomington and Indianapolis: Indiana University Press, 1990.

Navickas, Katrina. "The Search for 'General Ludd': The Mythology of Luddism." *Social History* 30, no. 3 (August 2005): 281–95.

Neal, Larry. "The Financial Crisis of 1825 and the Restructuring of the British Financial System." *Federal Reserve Bank of St. Louis Review*, June 1998. https://files.stlouisfed.org/files/htdocs/publications/review/98/05/9805ln.pdf.

Nenadic, Stena, and Sally Tuckett. *Colouring the Nation: The Turkey Red Printed Cotton Industry in Scotland c. 1840–1940*. Edinburgh: National Museums of Scotland, 2013.

Ó Gráda, Cormac. "Did Science Cause the Industrial Revolution?" University of Warwick, Working Paper Series, no. 205 (October 2014). www2.warwick.ac.uk/fac/soc/economics/research/centres/cage/manage/publications/205-2014_o_grada.pdf.

O'Brien, Patrick. "Agriculture and the Home Market for English Industry, 1660–1820." *English Historical Review* 100, no. 397 (October 1985): 773–800.

  "The Micro Foundations of Macro Invention: The Case of the Reverend Edmund Cartwright." *Textile History* 28, no. 2 (1997): 201–33.

O'Brien, Patrick, Trevor Griffiths, and Philip Hunt. "Political Components of the Industrial Revolution: Parliament and the English Cotton Textile Industry, 1660–1774." *Economic History Review* 44, no. 3 (August 1991): 395–423.

Ogilvie, Sheilagh. *Institutions and European Trade: Merchant Guilds, 1000–1800.* Cambridge and New York: Cambridge University Press, 2011.

State Corporatism and Proto-Industry: The Württemberg Black Forest, 1580–1797. Cambridge and New York: Cambridge University Press, 1997.

"The Economics of Guilds." *Journal of Economic Perspectives* 28, no. 4 (Fall 2014): 169–92.

"'Whatever Is, Is Right'? Economic Institutions in Pre-Industrial Europe." *Economic History Review* 60, no. 4 (November 2007): 649–84.

Olmstead, Alan L., and Paul W. Rhode. "Cotton, Slavery, and The New History of Capitalism." *Explorations in Economic History* 67, no. 1 (January 2018): 1–17.

Overton, Mark. *Agricultural Revolution in England: The Transformation of the Agrarian Economy 1500–1850.* Cambridge: Cambridge University Press, 1996.

Owen-Crocker, Gale R. "Brides, Donors, Traders: Imports into Anglo-Saxon England." In Angela Ling Huang and Carsten Jahnke (eds.), *Textiles and the Medieval Economy: Production, Trade, and Consumption of Textiles, 8th–16th Centuries,* 64–77. Oxford and Philadelphia: Oxbow Books, 2015.

Parkhill, John. *The History of Paisley.* Paisley: Robert Stewart, 1857.

Parthasarathi, Prasannan. *Why Europe Grew Rich and Asia Did Not: Global Economic Divergence, 1600–1850.* Cambridge and New York: Cambridge University Press, 2011.

Parthasarathi, Prasannan, and Ian Wendt, "Decline in Three Keys: Indian Cotton Manufacturing from the Late Eighteenth Century." In Giorgio Riello and Prasannan Parthasarathi, eds., *The Spinning World: A Global History of Cotton Textiles, 1200–1850,* 397–406. Oxford and New York: Pasold Research Fund and Oxford University Press, 2009.

Peck, Amelia, ed. *Interwoven Globe: The Worldwide Textile Trade, 1500–1800.* London: Thames and Hudson, 2013.

Peers, Douglas M. *India under Colonial Rule: 1700–1885.* London and New York: Routledge, 2013.

Phillips, Jonathan. *Holy Warriors: A Modern History of the Crusades.* New York: Random House, 2010.

Pollard, Sidney. "The Factory Village in the Industrial Revolution." *English Historical Review,* 79, no. 312 (July 1964): 513–31.

Pomeranz, Kenneth. *The Great Divergence: China, Europe, and the Making of the Modern World Economy.* Rev. ed. Princeton: Princeton University Press, 2000.

Power, Eileen E. "English Craft Gilds in the Middle Ages." *History,* New Series, 4, no. 16 (January 1920): 211–14.

Prakash, Om. "The English East India Company and India." In H. V. Bowen, Margarette Lincoln, and Nigel Rigby (eds.), *The Worlds of the East India Company,* 1–17. Woodbridge, Suffolk: The Boydell Press in association with the National Maritime Museum and the University of Leicester, 2002.

Prazniak, Roxann. "Siena on the Silk Roads: Ambrogio Lorenzetti and the Mongol Global Century, 1250–1350." *Journal of World History* 21, no. 2 (June 2010): 177–217.

Quataert, Donald, ed. *Consumption Studies and the History of the Ottoman Empire, 1550–1922: An Introduction.* Albany: State University of New York Press, 2000.

ed. *Manufacturing in the Ottoman Empire and Turkey, 1500–1950.* Albany: State University of New York, 1994.

Ransom, Philip John Greer. *The Victorian Railway and How It Evolved.* London: Heinemann, 1990.

Ray, Indrajit. *Bengal Industries and the British Industrial Revolution (1757–1857).* London and New York: Routledge, 2011.

Read, Donald. *Peterloo: The Massacre and Its Background.* Manchester: Manchester University Press, 1958.

Riello, Giorgio. "The Indian Apprenticeship: The Trade of Indian Textiles and the Making of European Cottons." In Giorgio Riello and Tirthankar Roy (eds.). *How India Clothed the World: The World of South Asian Textiles, 1500–1850,* 307–46. Leiden and Boston: Brill, 2009.

Cotton: The Fabric That Made the Modern World. Cambridge and New York: Cambridge University Press, 2013.

"Strategies and Boundaries: Subcontracting and the London Trades in the Long Eighteenth Century." *Enterprise and Society* 9, no. 2 (June 2008): 243–80.

Riello, Giorgio, and Prasannan Parthasarathi, eds. *The Spinning World: A Global History of Cotton Textiles, 1200–1850.* Oxford and New York: Pasold Research Fund and Oxford University Press, 2009.

Riello, Giorgio, and Tirthankar Roy, eds. *How India Clothed the World: The World of South Asian Textiles, 1500–1850.* Leiden and Boston: Brill, 2009.

Riley-Smith, Jonathan, ed. *The Oxford Illustrated History of the Crusades.* Oxford and New York: Oxford University Press, 1995.

Rimmer, W. G. *Marshalls of Leeds: Flax-Spinners, 1788–1886.* Cambridge: Cambridge University Press, 1960.

Rius. *Marx for Beginners.* New York: Pantheon, 2003.

Roberts, Lissa. "Producing (in) Europe and Asia, 1750–1850." *ISIS* 106, no. 4 (2015): 857–65.

Roberts, Lissa L., Simon Schaffer, and Peter Dear, eds. *The Mindful Hand: Inquiry and Invention from the Late Renaissance to Early Industrialisation.* Chicago: University of Chicago Press, 2008.

Rose, Mary B. *Firms, Networks, and Business Values: The British and American Cotton Industries since 1750.* Cambridge and New York: Cambridge University Press, 2000.

The Gregs of Quarry Bank Mill: The Rise and Decline of a Family Firm, 1750–1914. Cambridge and London: Cambridge University Press, 1986.

ed. *The Lancashire Cotton Industry: A History Since 1700.* Preston: Lancashire County Books, 1996.

Rosenband, Leonard N. *Papermaking in Eighteenth-Century France: Management, Labor, and Revolution at the Montgolfier Mill, 1761–1805.* Baltimore and London: Johns Hopkins University Press, 2000.

Rosenthal, Jean-Laurent, and R. Bin Wong. *Before and Beyond Divergence: The Politics of Economic Change in China and Europe.* Cambridge, MA: Harvard University Press, 2011.

Rule, John. *The Experience of Labour in Eighteenth-Century English Industry.* New York: St. Martin's Press, 1981.

Labouring Classes in Early Industrial England, 1750–1850. New York: Routledge, 2013. Orig. pub. 1986.

Russell, Andrew L., and Lee Vinsel. "After Innovation, Turn to Maintenance." *Technology and Culture* 59, no. 1 (January 2018): 1–25.

"Hail the Maintainers." *Aeon* 7 (April 2016). https://aeon.co/essays/innovation-is-overvalued-maintenance-often-matters-more.

Russell, Edmund. *Evolutionary History: Uniting History and Biology to Understand Life on Earth.* New York: Cambridge University Press, 2011.

Russell, Edmund, James Allison, Thomas Finger, John K. Brown, Brian Balogh, and W. Bernard Carlson. "The Nature of Power: Synthesizing the History of Technology and Environmental History." *Technology and Culture* 52, no. 2 (April 2011): 246–59.

Sabel, Charles, and Jonathan Zeitlin, eds. *World of Possibilities: Flexibility and Mass Production in Western Industrialization.* Cambridge: Cambridge University Press, 1997.

Salzmann, Ariel. "The Age of Tulips: Confluence and Conflict in Early Modern Consumer Culture (1550–1730)." In Donald Quataert (ed.), *Consumption Studies and the History of the Ottoman Empire, 1550–1922: An Introduction,* 83–106. Albany: State University of New York Press, 2000.

Sanderson, Michael. "Literacy and Social Mobility in the Industrial Revolution in England." *Past and Present* 56 (August 1972): 75–104.

Sardar, Marika. "Indian Textiles: Trade and Production." In Elizabeth A. Weinfield (ed.), *Heilbrunn Timeline of Art History.* New York: Metropolitan Museum of Art, 2000–2019. www.metmuseum.org/toah/hd/intx/hd_intx.htm (October 2003).

"Silk Along the Seas: Ottoman Turkey and Safavid Iran in the Global Textile Trade." In Amelia Peck (ed.), *Interwoven Globe: The Worldwide Textile Trade, 1500–1800,* 66–81. London: Thames and Hudson, 2013.

Scranton, Philip. *Proprietary Capitalism: The Textile Manufacture at Philadelphia, 1800–1885.* Cambridge and New York: Cambridge University Press, 1983.

Sen, Sudipta. *Empire of Free Trade: The East India Company and the Making of the Colonial Marketplace.* Philadelphia: University of Pennsylvania Press, 1998.

Seward, David. "The Wool Textile Industry 1750–1960." In J. Geraint Jenkins (ed.), *The Wool Textile Industry in Great Britain,* 34–48. London and Boston: Routledge & Kegan Paul, 1972.

Shammas, Carole. *The Pre-Industrial Consumer in England and America.* Oxford: Clarendon Press, 1990.

Sheppard, Eric. "Constructing Free Trade: From Manchester Boosterism to Global Management." *Transactions of the Institute of British Geographers* 30, no. 2 (June 2005): 151–72.

Sims, Richard. *Sailcloth, Webbing, and Shirts: The Crewkerne Textile Industry.* Bridport: Somerset Industrial Archaeological Society and the Gray Fund of the Somerset Archaeology and Natural History Society, 2015.

Singer, Charles, E. J. Holmyard, A. R. Hall, and Trevor Williams, eds. *A History of Technology.* 5 vols. New York and London: Oxford University Press, 1958.

Smail, John. *Merchants, Markets, and Manufacture: The English Wool Textile Industry in the Eighteenth Century.* New York: St. Martin's Press, 1999.

The Origins of Middle-Class Culture: Halifax, Yorkshire, 1660–1780. Ithaca and London: Cornell University Press, 1994.

Smiles, Samuel. *Industrial Biography: Iron Workers and Tool Makers.* London: John Murray, 1863.

The Lives of Boulton and Watt. London: John Murray, 1865.

Lives of the Engineers: The Locomotive, George and Robert Stephenson. London: John Murray, 1877.

Smith, Chloe Wigston. "'Callico Madams': Servants, Consumption, and the Calico Crisis." *Eighteenth-Century Life* 31, no. 2 (Spring 2007): 29–55.

Smith, Crosbie, Ian Higginson, and Phillip Wolstenholme. "'Avoiding Equally Extravagance and Parsimony': The Moral Economy of the Ocean Steamship." *Technology and Culture* 44, no. 3 (July 2003): 443–69.

Smith, Merritt Roe, and Leo Marx. *Does Technology Drive History? The Dilemma of Technological Determinism.* Cambridge, MA: MIT Press, 1994.

Smith, Pamela H. "Consumption and Credit: The Place of Alchemy in Johann Joachim Becher's Political Economy." In *Alchemy Revisited: Proceedings of the International Conference on the History of Alchemy at the University of Gronengen, 17–19 April 1989*, 215–28. Leiden and New York: E. J. Brill, 1990.

Smith, Pamela H., and Paula Findlen. *Merchants and Marvels: Commerce, Science, and Art in Early Modern Europe.* New York and London: Routledge, 2002.

Smithers, Henry. *Liverpool, Its Commerce, Statistics, and Institutions with a History of the Cotton Trade.* Liverpool: Thomas Kaye, 1825.

Staudenmaier, John M. *Technology's Storytellers: Reweaving the Human Fabric.* Cambridge, MA: MIT Press and the Society for the History of Technology, 1985.

Stearns, Peter N. *The Industrial Revolution in World History.* 4th ed. Boulder: Westview Press, 2012.

Steedman, Carolyn. *Labours Lost: Domestic Service and the Making of Modern England.* Cambridge and New York: Cambridge University Press, 2009.

Steffen, Will, Jacques Grinevald, Paul Crutzen, and John MacNeill. "The Anthropocene: Conceptual and Historical Perspectives." *Philosophical Transactions of the Royal Society* 369 (2011): 842–67.

Steffen, Will, Paul J. Crutzen, and John R. McNeill. "The Anthropocene: Are Humans Now Overwhelming the Great Forces of Nature?" *AMBIO. A Journal of the Human Environment* 36 (2007): 614–21.

Stevenson, John. *Popular Disturbances in England, 1700–1832.* New York: Routledge, 2013. Orig. pub. 1979.

Styles, John. "Fashion, Textiles and the Origins of Industrial Revolution." *East Asian Journal of British History*, 5 (March 2016): 161–89.

Thirsk, Joan. "Indian Cottons and European Fashion, 1400–1800." In Glenn Adamson, Giorgio Riello, and Sarah Teasley (eds.), *Global Design History*, 37–46. London: Routledge, 2011.

"Spinners and the Law: Regulating Yarn Standards in the English Worsted Industries, 1550–1800." *Textile History* 44, no. 2 (2013): 145–70.

The Dress of the People: Everyday Fashion in Eighteenth-Century England. New Haven and London: Yale University Press, 2007.

Subrahmanyam, Sanjay. *The Political Economy of Commerce: Southern India 1500–1650.* Cambridge and New York: Cambridge University Press, 1990.

Sugden, Keith. "An Occupational Study to Track the Rise of Adult Male Mule Spinning in Lancashire and Cheshire, 1777–1813." *Textile History* 48, no. 2 (2017): 160–75.

Sutcliffe, Thomas. *An Exposition of the Facts Relating to the Rise and Progress of the Woollen, Linen and Cotton Manufactures of Great Britain.* Manchester: P. Grant, 1843.

Sutton, Anne F. "Some Aspects of the Linen Trade c. 1130s to 1500, and the Part Played by the Mercers of London." *Textile History* 30, no. 2 (1999): 155–75.

Sweet, Rosemary. *The English Town, 1680–1840: Government, Society, and Culture.* New York: Pearson Education, 1999.

Swingen, Abigail L. *Competing Visions of Empire: Labor, Slavery, and the Origins of the British Atlantic Empire.* New Haven and London: Yale University Press, 2015.

Tatar, Maria. "Show and Tell: Sleeping Beauty as Verbal Icon and Seductive Story." *Marvels & Tales* 28, no. 1 (2014): 142–58.

Thirsk, Joan. "The Common Fields." *Past & Present* 29, no. 4 (December 1964): 3–25.

Economic Policy and Projects: The Development of a Consumer Society in Early Modern England. Oxford: Clarendon Press, 1978.

"The Origin of the Common Fields." *Past & Present* 33, no. 2 (April 1966): 142–47.

Thomas, Keith, et al. "Work and Leisure in Pre-Industrial Society (and Discussion)." *Past and Present* 29, no. 1 (December 1964): 29–66.

Thomis, Malcolm I. *The Luddites: Machine Breaking in Regency England.* Newton Abbot: David & Charles, 1970.

The Town Labourer and the Industrial Revolution. London: B. T. Batsford, 1974.

Thompson, E. P. *The Making of the English Working Class.* London: Victor Gollancz, 1964.

Thomson, James K. J. "Transferring the Spinning Jenny to Barcelona: An Apprenticeship in the Technology of the Industrial Revolution." *Textile History* 34, no. 1 (2003): 21–46.

Tilly, Charles. "State and Counterrevolution in France." *Social Research* 56, no. 1 (Spring 1989): 71–97.

Timmins, Geoffrey. *The Last Shift: The Decline of Handloom Weaving in Nineteenth-Century Lancashire.* Manchester and New York: Manchester University Press, 1993.

"Technological Change." In Mary B. Rose (ed.), *The Lancashire Cotton Industry: A History Since 1700*, 29–62. Preston: Lancashire County Books, 1996.

Made in Lancashire: A History of Regional Industrialisation. Manchester and New York: Manchester University Press, 1998.

"Roots of Industrial Revolution." In Alan Kidd and Terry Wyke (eds.), *Manchester: Making the Modern City*, 29–67. Liverpool: Liverpool University Press, 2016.

Toms, J. S., and Alice Katherine Shepherd. "Creative Accounting in the British Industrial Revolution: Cotton Manufacturers and the 'Ten Hours' Movement," *SSRN*, November 14, 2013. http://dx.doi.org/10.2139/ssrn.2354300.

Toynbee, Arnold. *Lectures on the Industrial Revolution in England: Popular Addresses, Notes, and Other Fragments*. London: Rivingtons, 1884.

Traill, Henry Duff. *Social England: A Record of the Progress of the People*. Vol. 6. New York and London: G.P. Putnam's Sons and Cassell & Company, 1897.

Trinder, Barrie. *The Industrial Revolution in Shropshire*. 3rd ed. Stroud, Gloucestershire: Phillimore & Co., 2016.

Tucker, Barbara M. *Samuel Slater and the Origins of the American Textile Industry, 1790–1860*. Ithaca and London: Cornell University Press, 1984.

Tunzelmann, G. N. von. *Steam Power and British Industrialization to 1860*. Oxford: Clarendon Press, 1978.

Turnbull, Geoffrey. *A History of the Calico Printing Industry of Great Britain*. Edited by John G. Turnbull. Altrincham: John Sherratt and Son, 1951.

Uglow, Jenny. *The Lunar Men: Five Friends Whose Curiosity Changed the World*. New York: Farrar, Straus and Giroux, 2002.

Unwin, George. *Samuel Oldknow and the Arkwrights: The Industrial Revolution at Stockport and Marple*. Manchester: Manchester University Press, 1968.

Usher, Abbott Payson. *A History of Mechanical Inventions*. Cambridge, MA: Harvard University Press, 1929.

Vella, Stephen. "Imagining Empire: Company, Crown and Bengal in the Formation of British Imperial Ideology, 1757–84." *Portuguese Studies* 16, no. 1 (2000): 276–97.

Vinsel, Lee. "Ninety-Five Theses on Innovation." Blogpost, November 12, 2015. Accessed February 9, 2019. http://leevinsel.com/blog/2015/11/12/95-theses-on-innovation.

Vries, Jan de. *The Industrious Revolution: Consumer Behavior and the Household Economy, 1650 to the Present*. Cambridge and New York: Cambridge University Press, 2008.

Wadsworth, Alfred P., and Julia De Lacy Mann. *The Cotton Trade and Industrial Lancashire, 1600–1780*. Manchester: Manchester University Press, 1931.

Walker, Michael John. "The Extent of the Guild Control of Trades in England, c. 1660–1820; A Study Based on a Sample of Provincial Towns and London Companies." Ph.D. Diss., University of Cambridge, 1985.

Wallerstein, Immanuel. *The Modern World-System I: Capitalist Agriculture and the Origins of the European World-Economy in the Sixteenth Century*. New York: Academic Press, 1974.

The Modern World-System II: Mercantilism and the Consolidation of the European World-Economy, 1600–1750. Berkeley and Los Angeles: University of California Press, 2011.

Walter, Dierk. *Colonial Violence: European Empires and the Use of Force.* Oxford and New York: Oxford University Press, 2016.

Walton, John K. *Blackpool.* Edinburgh: Edinburgh University Press, 1998.

Watt, Melinda. "Whims and Fancies: Europeans Respond to Textiles from the East." In Amelia Peck (ed.), *Interwoven Globe: The Worldwide Textile Trade, 1500–1800,* 82–103. London: Thames and Hudson, 2013.

Wendt, Ian C. "Four Centuries of Decline? Understanding the Changing Structure of the South Indian Textile Industry." In Giorgio Riello and Tirthankar Roy (eds.), *How India Clothed the World: The World of South Asian Textiles, 1500–1850,* 193–215. Leiden and Boston: Brill, 2009.

Watts, Martin. *Watermills.* Princes Risborough: Shire Publications, 2006.

Williams, Eric. *Capitalism and Slavery.* Chapel Hill: University of North Carolina Press, 1994. Orig. pub. 1944.

Williams, Mike, and with D. A. Farnie. *Cotton Mills in Greater Manchester.* Lancaster: Carnegie Publishing, 1992.

Wilson, R. G. *Gentlemen Merchants: The Merchant Community in Leeds, 1700–1830.* New York: Augustus M. Kelley and Manchester University Press, 1971.

"Gott, Benjamin (1762–1840)." In *Oxford Dictionary of National Biography,* 2004. http://dx.doi.org/10.1093/ref:odnb/49967.

Wise, M. Norton. "Measurement, Work, and Industry in Lord Kelvin's Britain." *Historical Studies in the Physical and Biological Sciences* 17, no. 3 (1986): 147–73.

"The Flow Analogy to Electricity and Magnetism, Part I: William Thomson's Reformulation of Action at a Distance." *Archive for History of Exact Sciences* 25, no. 1 (1981): 19–70.

Wolff, Janet, and John Seed, eds. *The Culture of Capital: Art, Power and the Nineteenth-Century Middle Class.* Manchester: Manchester University Press, 1988.

Wong, R. Bin. *China Transformed: Historical Change and the Limits of European Experience.* Ithaca and London: Cornell University Press, 1997.

Wood, Bertha. *Fresh Air and Fun: The Story of a Blackpool Holiday Camp.* Lancaster, UK: Palatine Books of Carnegie Publishing, 2005.

Wrigley, E. A. *Energy and the English Industrial Revolution.* Cambridge and New York: Cambridge University Press, 2010.

"The Growth of Population in Eighteenth-Century England: A Conundrum Resolved." *Past and Present* 98, no. 1 (February 1983): 121–50.

Zanier, Claudio. "Current Historical Research into the Silk Industry in Italy." *Textile History* 25, no. 1 (1994): 61–78.

Zimmerman, Andrew. *Alabama in Africa: Booker T. Washington, the German Empire, and the Globalization of the New South.* America in the World. Princeton and Oxford: Princeton University Press, 2010.

Zutshi, Chitralekha. "'Designed for Eternity': Kashmiri Shawls, Empire, and Cultures of Production and Consumption in Mid-Victorian Britain." *Journal of British Studies* 48, no. 2 (April 2009): 420–40.

Index

Pages in italics indicate an illustration.